Lecture Notes in Artificial Intelligence   10775

Subseries of Lecture Notes in Computer Science

LNAI Series Editors

Randy Goebel
    University of Alberta, Edmonton, Canada
Yuzuru Tanaka
    Hokkaido University, Sapporo, Japan
Wolfgang Wahlster
    DFKI and Saarland University, Saarbrücken, Germany

LNAI Founding Series Editor

Joerg Siekmann
    DFKI and Saarland University, Saarbrücken, Germany

More information about this series at http://www.springer.com/series/1244

Madalina Croitoru · Pierre Marquis
Sebastian Rudolph · Gem Stapleton (Eds.)

Graph Structures for Knowledge Representation and Reasoning

5th International Workshop, GKR 2017
Melbourne, VIC, Australia, August 21, 2017
Revised Selected Papers

Springer

Editors
Madalina Croitoru
LIRMM
Montpellier Cedex 5
France

Pierre Marquis
CRIL-CNRS and Université d'Artois
Lens
France

Sebastian Rudolph
Technische Universität Dresden
Dresden
Germany

Gem Stapleton
University of Brighton
Brighton
UK

ISSN 0302-9743   ISSN 1611-3349 (electronic)
Lecture Notes in Artificial Intelligence
ISBN 978-3-319-78101-3   ISBN 978-3-319-78102-0 (eBook)
https://doi.org/10.1007/978-3-319-78102-0

Library of Congress Control Number: 2018937369

LNCS Sublibrary: SL7 – Artificial Intelligence

Printed on acid-free paper

This Springer imprint is published by the registered company Springer International Publishing AG
part of Springer Nature
The registered company address is: Gewerbestrasse 11, 6330 Cham, Switzerland

Preface

Versatile and effective techniques for knowledge representation and reasoning (KRR) are essential for the development of successful intelligent systems. Many representatives of next-generation KRR systems are based on graph-based knowledge representation formalisms and leverage graph-theoretical notions and results. The goal of the workshop series on Graph Structures for Knowledge Representation and Reasoning (GKR) is to bring together the researchers involved in the development and application of graph-based knowledge representation formalisms and reasoning techniques.

This volume contains extended and revised selected papers of the fifth edition of GKR, which took place in Melbourne, Australia, on August 21, 2017. Like the previous editions, held in Pasadena, USA (2009), Barcelona, Spain (2011), Beijing, China (2013), and Buenos Aires, Argentina (2015), the workshop was associated with IJCAI (the International Joint Conference on Artificial Intelligence), thus providing the perfect venue for a rich and valuable exchange. Beside the extended workshop papers, this volume also contains two invited contributions of core GKR community members.

The scientific program of this workshop included many topics related to graph-based knowledge representation and reasoning such as argumentation, conceptual graphs, formal concept analysis, graphical models, Bayesian networks, concept diagrams, and many more. All in all, the fifth edition of the GKR workshop was very successful. The papers coming from diverse fields all addressed various issues in knowledge representation and reasoning and the common graph-theoretic background allowed us to bridge the gap between the different communities. This made it possible for the participants to gain new insights and inspiration.

We are grateful for the support of IJCAI and we would also like to thank the Program Committee of the workshop for their hard work in reviewing papers and providing valuable guidance to the contributors. But, of course, GKR 2017 would not have been possible without the dedicated involvement of the contributing authors and participants.

February 2018

Madalina Croitoru
Pierre Marquis
Sebastian Rudolph
Gem Stapleton

Organization

Workshop Chairs

Madalina Croitoru   LIRMM, Université Montpellier II, France
Pierre Marquis   CRIL-CNRS and Université d'Artois, France
Sebastian Rudolph   Technische Universität Dresden, Germany
Gem Stapleton   University of Brighton, UK

Program Committee

Simon Andrews   Sheffield Hallam University, UK
Abdallah Arioua   INRA, LIRMM, Université Montpellier II, France
Zied Bouraoui   Cardiff University, UK
Dan Corbett   Optimodal Technologies, USA
Olivier Corby   Inria, France
Cornelius Croitoru   University Al.I.Cuza Iaşi, Romania
Frithjof Dau   SAP, Germany
Juliette Dibie-Barthélemy   AgroParisTech, France
Peter Eklund   IT University of Copenhagen, Denmark
Catherine Faron Zucker   Université Nice Sophia Antipolis, France
Sebastien Ferre   Université de Rennes 1, France
Christophe Gonzales   LIP6, Université Paris 6, France
Ollivier Haemmerlé   IRIT, Université Toulouse le Mirail, France
John Howse   University of Brighton, UK
Bernard Moulin   Université Laval, Canada
Laura Papaleo   Université Paris-Sud, France
Simon Polovina   Sheffield Hallam University, UK
Uta Priss   Ostfalia University, Germany
Karim Tabia   Université d'Artois, France
Srdjan Vesic   CRIL, CNRS – Université d'Artois, France
Nic Wilson   Insight UCC, Cork, Ireland
Stefan Woltran   Vienna University of Technology, Austria
Bruno Yun   Université Montpellier II, France

Additional Reviewers

Jan Maly   Vienna University of Technology, Austria
Axel Polleres   Vienna University of Economics and Business, Austria

Contents

Extended Workshop Papers

Exploring, Reasoning with and Validating Directed Graphs by Applying Formal Concept Analysis to Conceptual Graphs

Simon Andrews and Simon Polovina[✉]

Conceptual Structures Research Group, Department of Computing, Communication
and Computing Research Centre, Sheffield Hallam University, Sheffield, UK
{s.andrews,s.polovina}@shu.ac.uk

Abstract. Although tools exist to aid practitioners in the construction
of directed graphs typified by Conceptual Graphs (CGs), it is still quite
possible for them to draw the wrong model, mistakenly or otherwise.
In larger or more complex CGs it is furthermore often difficult–without
close inspection–to see clearly the key features of the model. This paper
thereby presents a formal method, based on the exploitation of CGs as
directed graphs and the application of Formal Concept Analysis (FCA).
FCA elucidates key features of CGs such as pathways and dependencies,
inputs and outputs, cycles, and joins. The practitioner is consequently
empowered in exploring, reasoning with and validating their real-world
models.

1 Introduction

A directed graph–or "digraph"–is a graph whose edges have direction and are
called arcs [9,11]. Arrows on the arcs are used to encode the directional information: an arc from vertex A to vertex B indicates that one may move from A
to B but not from B to A. Such graphs for example are used in computer science
as a representation of the paths that might be traversed through a program, or
in higher-level conceptual models where concepts are related to each other by
relations that gain additional semantics (i.e. meaning) by defining the direction
between the source and target concepts. A classic illustration is a cat that sits
on a mat [18]. In this simple example 'sits-on' is the semantic relation where the
direction goes from cat to mat and not vice versa.

CGs (Conceptual Graphs) are digraphs that enable modellers to express
meaning in a form that is logically precise whilst being humanly readable, and
serve as an intermediate language for translating between computer-oriented
formalisms and natural languages [14,19]. CGs graphical representation thereby
serve as a readable, but formal specification language for systems design or other
practitioners using this approach [10]. CGs are however drawn by hand. Tools
such as CoGui and CharGer already exist to assist the practitioner in creating
a well-formed CG (Conceptual Graph) that adheres to the prescribed grammar
and syntax [1,2]. However there is no guarantee that a model created using CGs

© Springer International Publishing AG, part of Springer Nature 2018
M. Croitoru et al. (Eds.): GKR 2017, LNAI 10775, pp. 3–28, 2018.
https://doi.org/10.1007/978-3-319-78102-0_1

is correct in terms of its validity. The modeller may have a misconception of the system being modelled or may simply make mistakes in its construction–things that still conform to the syntax and grammar but result in an invalid model.

It can be difficult to explore and validate a large and complex CG by inspection. It is this problem that this paper addresses by providing an automated method whereby key features of CGs are captured, reported and visualised. The modeller would thus be assisted in exploring and validating their CGs. The method makes use of the inherent direction of Concept-relation-Concept triples in CGs to transform these triples into binary relations and thus expose them to Formal Concept Analysis (FCA) [8]. The process is automated in a tool called *CGFCA* and has two stages; firstly parsing a CG file (in the ISO common logic *cgif* format [19]) to extract the CG triples and secondly, converting these triples into corresponding binary relations that accentuate the directed pathways in the original CG, as described next in Sect. 2. The triples-to-binaries function is carried out using an implementation of the *Triples2Binaries* algorithm, specifically described in Subsect. 2.2.

2   Transforming CG Digraphs: Triples into Binary Relations

If triples are extracted from a CG in the form *Source Concept → relation → Target Concept*, each such triple can easily be represented as a corresponding binary relation i.e. *Source Concept-relation, Target Concept*. Where the Target Concept then becomes a Source Concept for a following relation, this can be captured in additional binary relations, where the original Source Concept-relation is paired with subsequent Target Concepts. To illustrate the source-target structure, Fig. 1 shows a simple CG with the CG Concepts, [Cat], [Mat] and [Colour: Grey]. [Cat] and [Mat] are linked by the CG relation (sits-on) and [Mat], [Colour: Grey] are linked by (has-attribute). (In simple English, the CG describes a cat that sits on a grey mat.) We can say that the

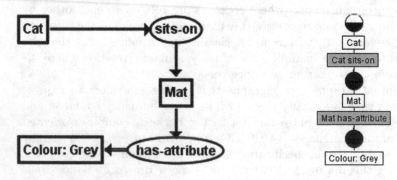

Fig. 1. Simple CG   **Fig. 2.** FCL for simple CG

target Concept [Mat] is dependent on the source Concept-relation pair [Cat] →
(sits-on) and the target Concept [Colour:Grey] is dependent on its source
Concept-relation pair [Mat] → (has-attribute) (or alternatively, the source
Concept-relation pair [Cat] → (sits-on) results in the target Concept [Mat]
and the source Concept-relation pair [Mat] → (hasattribute) results in the target Concept [Colour: Grey]). The CG triple ([Cat], (sits-on), [Mat]) can
be converted into the binary relation ([Cat]-(sits-on), [Mat]). Likewise the
CG triple ([Mat], (hasattribute), [Colour: Grey]) can be converted into the
binary relation ([Mat](has-attribute), [Colour: Grey]).

There is also a binary relation between [Cat] and [Colour: Grey] indirectly
through [Cat]-(sits-on). Hence [Colour: Grey] also depends (indirectly) on
[Cat], which is of course sitting on that mat.

Simple CG	Cat sits-on	Mat has-attribute
Cat		
Mat	×	
Colour : Grey	×	×

Fig. 3. The simple CG as a cross-table

The set of binary relations can be simply represented in a cross-table and
Fig. 3 shows the corresponding cross table for this simple example, with rows
representing CG Concepts and columns CG Source Concept-relations. The cross-
table is known as a Formal Context in FCA, so by converting CGs into these
binary relations, FCA can then be applied. Figure 2 displays the resulting Formal
Concept Lattice (FCL). This approach was derived after we compared it with
Wille's mapping of CGs to FCA ('Concept Graphs') in an earlier study [5,20].
Figures 4, 5, and 6 show the CG, FCL and cross-table (Formal Context) for a
larger CG using the same Cat on Mat example. Figures 7, 8, and 9 show the CG,
FCL and cross-table for a further extended version of this example. This time it
has two input CG Concepts, [Cat: Gwyn] and [Cat: Bumbles] thus depicting
the specific cats Gwyn and Bumbles as the respective CG referent for each CG
Concept as shown.

2.1 Obtaining Triples from a Conceptual Graph: A Parser for *cgif*

To automate this process, a parser was created that operates on the standard
CG file format, *cgif*. To illustrate the format, below is the *cgif* for the 'Cats on
the Mat' CG in Fig. 7:

```
[Material: Fleece] [Cat: Gwyn] [Mat: *x1] [Cat: Bumbles] [Colour: Grey]
(sits-on Bumbles Gwyn ?x1)(has-attribute Fleece Grey)(made-from ?x1 Fleece)
```

The first line in the *cgif* defines the CG Concepts and the second line defines
the CG relations. In line with CGs theory where the referent is unknown, *cgif*

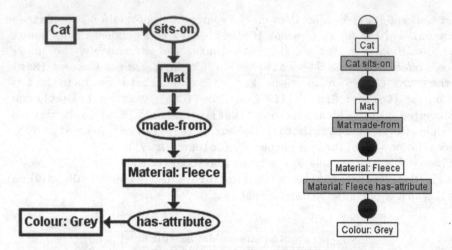

Fig. 4. CG with 4 concepts   **Fig. 5.** FCL for CG with 4 concepts

With 4 Concepts	Cat sits-on	Mat made-from	Material: Fleece has-attribute
Cat			
Mat	×		
Material : Fleece	×	×	
Colour : Grey	×	×	×

Fig. 6. The 4 concept CG as a cross-table

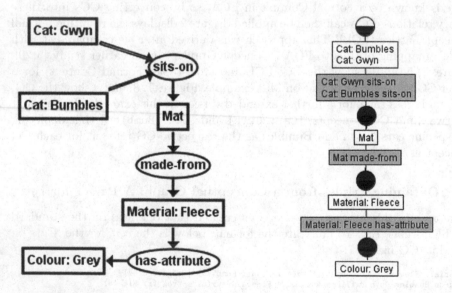

Fig. 7. CG with 2 input concepts   **Fig. 8.** FCL for 2 input concepts

2 Input Concepts	Cat: Bumbles sits-on	Cat: Gwyn sits-on	Mat made-from	Material: Fleece has-attribute
Cat : Bumbles				
Cat : Gwyn				
Mat	×	×		
Material : Fleece	×	×	×	
Colour : Grey	×	×	×	×

Fig. 9. The 2 input CG as a cross-table

uses generic referents such as x1 and x2, with a preceding * (in CG Concepts) or ? (in CG relations). Each relation is defined in the *cgif* with a list of referents comprising one or more source CG Concept referents and a target CG Concept referent. The final referent in the list is always the target referent. Thus, in the relation made-from, x1 is the source and Fleece is the target, and in the relation sits-on, Bumbles and Gywn are sources and x1 is the target.

The parser first extracts the CG Concepts from the *cgif*, creating an integer index for each CG Concept and separating the type labels and referents (see Table 1a). The parser then extracts the CG relations from the *cgif*, creating an integer index for each CG relation and separating the type labels and lists of referents (see Table 1b).

Table 1. Information extracted by parser from Cats on the Mat *cgif*

No.	Label	Referent
1	Material	Fleece
2	Cat	Gwyn
3	Mat	x1
4	Cat	Bumbles
5	Colour	Grey

(a) CG Concepts

No.	Label	Referents
1	sits-on	Bumbles Gwyn x1
2	has-attribute	Fleece Grey
3	made-from	x1 Fleece

(b) CG relations

If there are co-referent CG Concepts or relations, the parser will form the corresponding joins. For CG Concepts, as each Concept label and referent is extracted from the *cgif*, the referent is compared to the list of Concept referents already extracted. If a match is found, instead of adding a new Concept, the parser compares the two Concept labels. If they are different, it concatenates the new label with the existing label in the list, if not the parser simply moves on to the next Concept in the *cgif*. A similar process is carried when parsing the CG relations in the *cgif*, but here it is the list of referents associated with the relation that is compared: for two CG relations to be co-referent they must have the same sources and target. For examples of joining co-referents see Sect. 4.8.

Once the CG Concepts and relations have been extracted (and any co-referent joins made), the parser then uses the referents for each relation to create corresponding triples by looking up the index number of the relation's source and target CG Concepts corresponding to the relation's referents. Table 2 contains the triples created from Table 1. The triples thus created are now ready for the process of converting them to corresponding binary relations.

Table 2. Cats on the Mat triples

Source	Relation	Target
4	1	3
2	1	3
1	2	5
3	3	1

2.2   A Triples-to-Binaries Algorithm

Figure 10 is an algorithm, *Triples2Binaries*, that along with its subroutine *AddBinary* (Fig. 11), converts a set of triples, T, into a corresponding set of binaries, B, exploiting the direction in the triples as explained above. It is a generalised form of the *CGtoFCA* algorithm previously presented [5]. Whilst its application to CGs is the focus of this paper, the more general form makes it applicable to directed triples obtained from any source, including UML, RDF, OWL, the Entity-Relation Diagram and linked data. *Triples2Binaries* also includes some refinements not present in *CGtoFCA*, namely; the ability to detect 'direct pathways' and cycles in a CG. A direct pathway through a CG is a path from an input CG Concept to an output CG Concept, where an input CG Concept is one with no edges entering it and an output CG Concept is one with no edges leaving it. Features such as direct pathways and cycles often have significant meaning in a CG but are not always easily apparent (particularly in large CGs). The main algorithm, *Triples2Binaries*, simply iterates through the set of triples, T, sending each triple, (s, r, t) to the subroutine *AddBinary*. In (s, r, t), s denotes the *source*, r denotes the *relation* and t denotes the *target*. Each triple enumerated in *Triples2Binaries* will be the start of a new pathway. A pathway is recorded by *AddBinary* as a set of (*source*, *relation*) pairs in *path*.

AddBinary takes each triple (s, r, t), adds (s, r) to the current path (line 2) and then adds the corresponding binary $((s, r), t)$ to the set of binaries, B (line 3).

Line 4 is a test for detecting a direct pathway in the CG: if the current source, s, is an input CG Concept and the current target, t, is an output CG Concept, then there is a direct pathway from s to t. In which case, the current path along with t is recorded as a direct pathway.

Line 6 is the condition for detecting a cycle in the CG: if the current source, s, is the same as the current target, t, there is a cycle, recorded in line 7 as the current path along with the current target.

Line 8 defines the terminating condition for recursion (thus preventing infinite loops around cycles): if the current target, t, is already in the current path then *AddBinary* terminates. Otherwise, line 9 iterates through the set of triples, T, to test for links (line 10): if the current target, t, also appears as a source, i, in the set of triples, *AddBinary* is called recursively (line 11), passing the current source, s, the current relation, r, and the new target, k.

Note that the condition for a cycle (line 6) cannot be used as the terminating condition for recursion. This is because the starting point for a cycle can occur at any point in a pathway. A pathway begins with the source, s, and if the starting point of a cycle begins later than s, then s will never equal t and we would have an infinite loop around that cycle.

```
1  begin
2  |   path ← ∅
3  |   foreach (s, r, t) ∈ T do
4  |   └   AddBinary(s, r, t, path)
5  end
```

Fig. 10. *Triples2Binaries(T)*

3   The CGFCA Tool

The *cgif* parser and *Triples2Binaries* algorithm were implemented together to form a software tool called CGFCA. The architecture of CGFCA is shown in Fig. 12. The *cgif* parser inputs a CG in the form of a *cgif* file and creates a corresponding set of CG (source Concept, relation, target Concept) triples as described in Sect. 2.1. The triples are then passed to Triples to Binaries which converts them into ((source Concept, relation), target Concept) binaries, including the computation of all binaries with indirect target Concepts, as described in Sect. 2.2. Triples to Binaries also carries out an analysis of the CG and reports the following features: input Concepts, output Concepts, direct pathways (from an input Concept to an output Concept), cycles and pathways with multiple routes (these are multiple pathways from the same input Concept-relation to the same output Concept). These multiple routes were considered worth detecting and reporting as they may indicate redundant pathways in a CG. The ((source Concept, relation), target Concept) binaries computed by Triples to Binaries

```
 1  begin
 2  │   path ← path ∪ {(s, r)}
 3  │   B ← B ∪ {((s, r), t)}
 4  │   if IsInput(s) and IsOutput(t) then
 5  │   └   RecordDirectPathway(path, t)
 6  │   if s = t then
 7  │   └   RecordCycle(path, t)
 8  │   if ¬ ∃(x, y) ∈ path | t = x then
 9  │   │   foreach (i, j, k) ∈ T do
10  │   │   │   if t = i then
11  │   │   └   └   AddBinary(s, r, k, path)
12  end
```

Fig. 11. *AddBinary*(*s*, *r*, *t*, *path*)

are then passed to a simple Formal Context Creator where the (source Concept, relation) in each binary is treated as a formal attribute and each target Concept is treated as a formal object. The formal context is output in the standard *cxt* format for FCA.

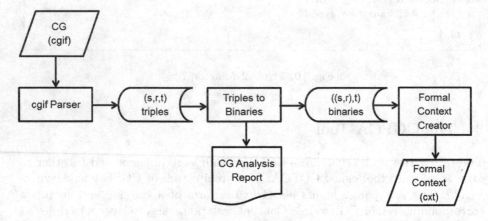

Fig. 12. CGFCA architecture

The formal context output by *CGFCA* can then be visualised as a Formal Concept Lattice (FCL) using an appropriate tool, such as *ConceptExplorer* (*ConExp*)[1] or as a Formal Concept Tree using In-Close [3,4]. Such visualisations clearly highlight further CG features such as cycles and co-referent joins.

---

[1] http://conexp.sourceforge.net/.

4 Highlighting Key Features of a CG Using CGFCA

This Section uses simple CG examples to illustrate the use of the GCFCA tool in detecting and highlighting features of CGs and how corresponding FCLs allow a GC to be explored in a formal, hierarchical, visualisation.

4.1 Paths and Dependencies

Figures 13 and 14 respectively illustrate the CG and FCL for the dependencies described earlier in a larger example–as well as two paths–between the source Concept [Person: Simon] and the target Concept [City: London]. As well as the intermediate target Concepts that in turn become source Concepts (i.e. [Coach: #564] and [Hotel: OpenSky]), this example shows CG referents, namely Simon, #564, OpenSky and London. ([Colour: Grey] from Fig. 2 was also a CG Concept with a referent.) The referents are instances of their respective type label in the CG Concept e.g. London is a referent of the type label City, and #564 the numeric identifier for a Coach that in the context of Fig. 13 could be read as the number of the coach that goes to London. In addition to the direct dependencies such as [Hotel: OpenSky]) on [Person: Simon]-(books) there are indirect dependencies detected in accordance with *AddBinary* line 4 described earlier In Subsect. 2.2. These are: (a) [City: London]) on [Person: Simon]-(books), and (b) [Person: Simon]-(travels-to) through the other path that has the intermediate Concept [Coach: #564]. The starting (or input) Concepts and ending (or output) Concepts are usefully reported by the *CGFCA* software i.e. Inputs: "Person: Simon". Outputs: "City: London". The output also states: Direct Pathway: Person: Simon - books - Hotel: OpenSky - location - City: London and Direct Pathway: Person: Simon - travels-by - Coach: #564 - destination - City: London.

Fig. 13. Paths and dependencies CG

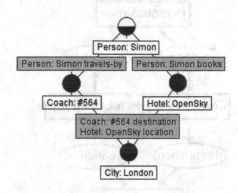

Fig. 14. Paths and dependencies FCL

In simple terms, Simon's trip to London depends on travelling there by coach and booking into the OpenSky hotel. Of course in this still-simple example this knowledge can be gleaned from the CG alone thereby obviating the need for *CGFCA*. However it is more likely that these patterns will appear in larger CGs where it is not so evident, perhaps unknowingly as they are drawn by hand and obfuscated by the size of the larger model. *CGFCA* and the consequent computer-generated FCL will highlight within such digraphs the 'diamond' looking patterns that represent multiple pathways thus alerting their existence–hence validity–to the modeller.

4.2   Cycles

It is natural that digraphs may contain one or more cycles. Figures 15 and 16 respectively illustrate an example of a CG and FCL that is a cycle. Note that this example is similar to the previous paths and dependencies example in Figs. 13 and 14. This time the direction of the hotel booking path goes in the opposite direction, thus creating the cycle. The renaming of the relations i.e. `location` to `location-of` and `books` to `booked-by` correctly reflect the new direction. It is common however to name or use relations that cause cycles to occur inadvertently such as possibly in this example. A cycle may of course be desired, but the modeller will in any event be alerted to its validity by the FCL (here Fig. 16) in accordance with *AddBinary* line 7 described earlier in Subsect. 2.2. The *CGFCA* output highlights why the Fig. 16 lattice looks as it does: There are no inputs. There are no outputs. Cycle: City: London - location-of - Hotel: OpenSky - booked-by - Person: Simon - travels-by - Coach: #564 - destination - City: London. Every Concept is dependent on all the other Concepts with no hierarchy, thus they become grouped together in the FCL.

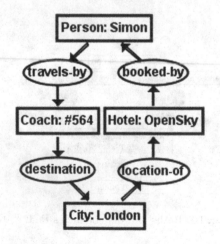

Fig. 15. CG that is a cycle

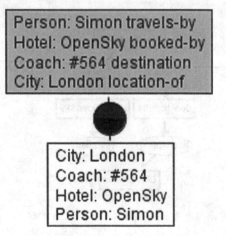

Fig. 16. FCL of cycle

4.3   Joins

Figures 17 and 18 respectively illustrate the CG and FCL for Concepts that are co-referent. Co-referents occur when Concepts have the same referent, which in this case is Gywn in Pet and Cat. Where a source and target Concept are directly linked by more than one relation, the associated relations are in effect co-referent. This behaviour is highlighted by Figs. 19 and 20.

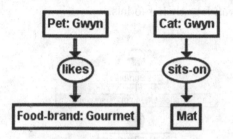

Fig. 17. CG with co-referent concept

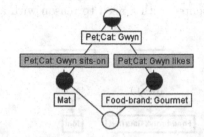

Fig. 18. FCL with co-referent concept

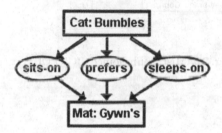

Fig. 19. CG with co-referent relations

Fig. 20. FCL with co-referent relations

Before the *Triples2Binaries* algorithm in *CGFCA* is called, the *CGFCA cgif* parser detects co-referent CG Concepts and co-referent CG relations and because they refer to the same object or instance it joins the Concepts and relations automatically (see Sect. 2.1). Furthermore it concatenates the Concept type or relation labels, using ';' as the delimiter.

The outcome is evident in the FCL for Figs. 18 (i.e. Pet;Cat) and 20 (i.e. sleeps-on;sits-on;prefers). This approach is akin to the maximal common subtype in CGs (or intersection); thus Gywn is (a) a Pet Cat, and (b) sleeps, sits on, and likes the Mat[2].

A common error (particularly in larger or more complex models) is to give different types the same referent by mistake. Take for example the CG Fig. 19.

---

[2] Note Mat here has a latent referent, in accordance with CGs theory; hence we can simply refer to it through the definite article 'the'.

In that Figure let's change [Mat: Gwyn's] to [Mat: Gwyn], assuming that it was mistyped by the modeller in the first place. As a result, [Mat: Gwyn] will inadvertently join with the [Cat: Gwyn] and [Pet: Gwyn] CGs from Fig. 17. Figure 21 shows the CGs for this scenario including the mistake, and Fig. 22 demonstrates the result. Now Gwyn is not only a Pet Cat but a Mat too! And Bumbles sleeps-on, sits-on and prefers Gwyn as a Mat (rather than Gwyn's Mat) while Gwyn sits on another Mat, all of which is nonsensical as the FCL reveals. Like the previous pathways and cycles examples, the practitioner is immediately presented with a need to reason with and validate their models.

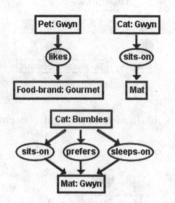

Fig. 21. CG with co-referents

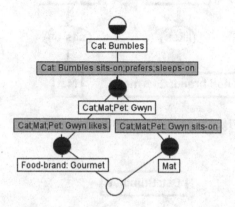

Fig. 22. Mistakenly joined CGs FCL

4.4   n-adic

Apart from Fig. 7 earlier, the CG relations so far have been *2-adic* i.e. only one source CG Concept pointing to the relation. *2-adic* CG relations are also known as *dyadic* CG relations. A CG relation may however have more than one source CG Concept; hence an *n*-adic CG relation has *n* source CG Concepts. The CG relation sits-on in Fig. 7 is *3-adic*, or *triadic*.

Figures 23 and 25 highlights the relation sits-on being stated as being dyadic or triadic respectively. CG relations may any number of source CG

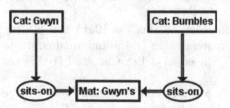

Fig. 23. CG with 2-adic relation

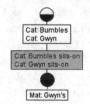

Fig. 24. FCL with 2-adic relation

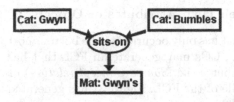

Fig. 25. CG with 3-adic relation   **Fig. 26.** FCL with 3-adic relation

Concepts pointing to them[3]. Figures 24 and 26 reveal that the FCL for Figs. 23 and 25 turn out to be identical, thus two representations of the same meaning. Unsurprisingly, the *CGFCA* output is identical for both the 2-adic and the 3-adic: Inputs: "Cat: Gwyn" "Cat: Bumbles" Outputs: "Mat: Gwyn's" Direct Pathway: Cat: Gwyn - sits-on - Mat: Gwyn's Direct Pathway: Cat: Bumbles - sits-on - Mat: Gwyn's.

Certain CG relations such as '(share)' inherently can only have certain n-adic values. For the share case, there need to be two or more things to have something shared between them, hence share has to be at least triadic i.e. $\geq$ 3-adic. As *CGFCA* would provide the same outcome even if the share CG relation was modelled as dyadic accidentally by the modeller, it would still be correctly depicted in the FCL. For completeness, Figs. 27, 28, 29 and 30 respectively demonstrate this outcome.

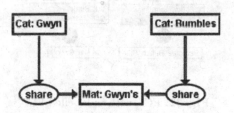

Fig. 27. CG, 'wrong' 2-adic share   **Fig. 28.** FCL, 'corrected' share

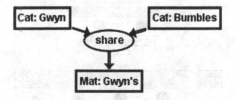

Fig. 29. CG, 'correct' 3-adic share   **Fig. 30.** FCL, 'correct' 3-adic share

---

[3] CG relations may however have only one target CG Concept [17].

4.5   Formal Concepts Without Their Own Attributes or Objects

Unlike the examples shown thus far where it has only occurred at the bottommost (or *infimum*) Formal Concept in an FCL, CGs may generate an FCL that has Formal Concepts without their own attributes (i.e. *Source Concept-relation*) or objects (i.e. *Target Concept*) in the middle of the FCL. Figure 31 has generated such a formal concept as evident in Fig. 32.

Essentially this is because [Cat: Bumbles] and [Cat: Gywn] *both* sit-on the [Mat: Gwyn's] *and* have the heritage of [Pedigree: British_Blue]. This pattern can be gleaned from the corresponding *CGFCA* output for Figs. 31 and 32:

Inputs: "Cat: Bessie" Outputs: "Mat" "Pedigree: British_Blue"
Direct Pathway: Cat: Bessie - offspring-of - Cat: Bumbles - heritage - Pedigree: British_Blue
Direct Pathway: Cat: Bessie - offspring-of - Cat: Bumbles - sits-on - Mat
Direct Pathway: Cat: Bessie - offspring-of - Cat: Gwyn - sits-on - Mat
Direct Pathway: Cat: Bessie - offspring-of - Cat: Gwyn - heritage - Pedigree: British_Blue

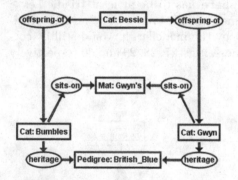

Fig. 31. CG leading to unlabelled FC

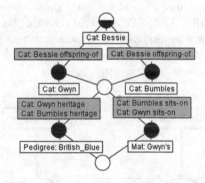

Fig. 32. FCL with unlabelled FC

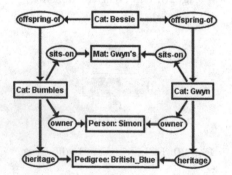

Fig. 33. Larger CG, unlabelled FC

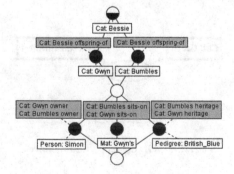

Fig. 34. Larger FCL, unlabelled FC

The only way to traverse the FCL to capture these relations is through the unlabelled Formal Concept in between.

Figures 33 and 34 evidence the pattern in a larger example where, essentially, [Cat: Bumbles] and [Cat: Gywn] *both* sit-on the [Mat: Gwyn's] *and* have the heritage of [Pedigree: British_Blue], *and* have as their owner the [Person: Simon]. The *CGFCA* output underpins the pattern:

Inputs: "Cat: Bessie" Outputs: "Person: Simon" "Mat: Gwyn's" "Pedigree: British_Blue"

Direct Pathway: Cat: Bessie - offspring-of - Cat: Bumbles - owner - Person: Simon

Direct Pathway: Cat: Bessie - offspring-of - Cat: Bumbles - heritage - Pedigree: British_Blue

Direct Pathway: Cat: Bessie - offspring-of - Cat: Bumbles - sits-on - Mat: Gwyn's

Direct Pathway: Cat: Bessie - offspring-of - Cat: Gwyn - heritage - Pedigree: British_Blue

Direct Pathway: Cat: Bessie - offspring-of - Cat: Gwyn - owner - Person: Simon

Direct Pathway: Cat: Bessie - offspring-of - Cat: Gwyn - sits-on - Mat: Gwyn's.

4.6 Further Exploring *n*-adity

For Fig. 33 we can also identify the presence of 3-adic (triadic) relations, as CG Fig. 35 reveals. Note once more that the FCL Fig. 36 is identical to Fig. 34.

Figure 37 has a CG with heritage as a *4*-adic relation, essentially adding that [Cat: Bessie] has the heritage of [Pedigree: British_Blue] too, along with [Cat: Bumbles] and [Cat: Gywn]. Through the unlabelled Formal Concept the 4th adic is highlighted by Fig. 38.

4.7 Further Exploring Co-Referent Links

Figure 37's CG can be restated using a co-referent link as shown by Fig. 39. In this Figure, the CG Concept [Pedigree: British_Blue] appears twice. Note also that the 4-adic heritage relation has disappeared, or so it would appear?

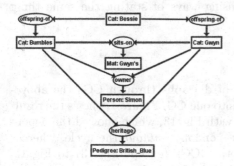

Fig. 35. Same CG, 3-adic

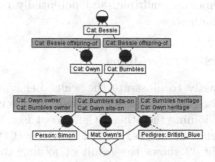

Fig. 36. Resulting same FCL

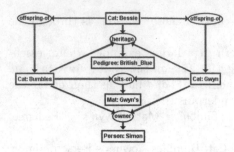
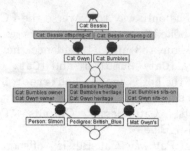

Fig. 37. CG, with 4-adic   **Fig. 38.** FCL, with 4-adic

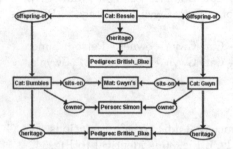
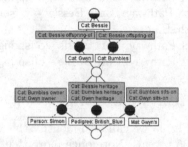

Fig. 39. Same CG, but co-referent   **Fig. 40.** Resulting same FCL

Note that Fig. 40, which is the FCL for Fig. 39 is identical to the FCL Fig. 38.
The *CGFCA* parser applies the CG join operation as before thus causing the
co-referents–as they are the same CG referent–to be joined [14,17,19]. The significance of this example is that it reminds us that CGs may be hand-drawn
in different ways (e.g. different adity, or using co-referents advertently–or inadvertently as we saw with [Mat: Gwyn] in CG Fig. 21 and the corresponding
FCL Fig. 22). However the FCL will represent them in *one* way, thus potentially
removing multiple, and potentially confusing ways of stating the same thing
differently.

4.8   Larger Joins

Lastly to illustrate the wider behaviour of digraphs through CGs the above-
discussed examples are essentially joined into one CG. Figure 41 shows the result
of joining the other CGs (except Fig. 15) with Fig. 13, which showed the dependency from [Person: Simon] to [City: London] *without* the cycle, whereas
Fig. 42 shows the result of joining the other CGs (except Fig. 13) to Fig. 15,
which showed [Person: Simon] to (and from) [City: London] *with* the cycle.

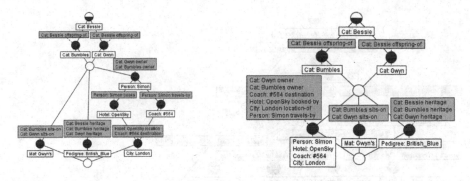

Fig. 41. Joined FCL no cycle   **Fig. 42.** Joined FCL with cycle

5   A Realistic Example

The simple but expressive examples presented thus far demonstrate how digraphs can be explored and validated through *Triples2Binaries* as exemplified by *CGFCA*. Previous work has indicated *CGFCA*'s value in the business information systems modelling domain [15]. Based on an example from that work, a more comprehensive example is now presented from that real-world domain. Whilst the example uses the terminology of that domain, this example will be explained such that it can be more widely understood.

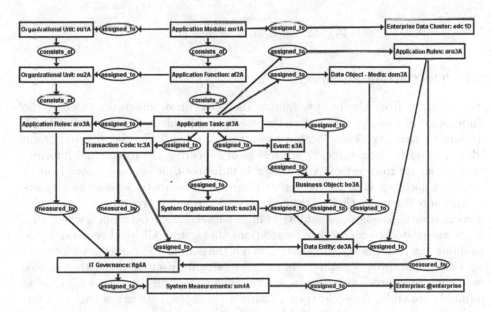

Fig. 43. Application module CG

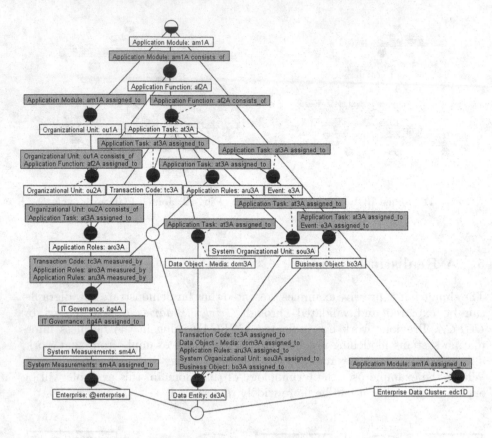

Fig. 44. Application module FCL

5.1   The Current Situation

As explained from the outset, human modellers draw diagrams to elicit the dimensions of some problem that becomes too difficult to understand through discursive narrative alone. We have seen that through CGs, the directed graph (digraph) offers the significant advantage of capturing the directional information i.e. an arc from vertex A to vertex B indicates that one may move from A to B but not from B to A. As well as their use in computer science as a representation of the paths that might be traversed through a program, the examples demonstrate digraphs' applicability in higher-level conceptual models where concepts are related to each other by relations that gain additional semantics (i.e. meaning) by defining the direction between the source and target concepts.

CGs (Conceptual Graphs) are an expressive form of digraphs that enable modellers to express meaning in a form that is logically precise whilst being humanly readable. As such, they provide a conceptual structure that can formally describe the given problem being modelled. CGs, in common with many other forms of diagrams are however drawn by hand, even with the assistance of software tools such as CoGui suggested earlier [2]. Currently, the modeller enters the digraphs–in this case CGs–into the tool manually and relies on the tool to work with the potentially erroneous CGs entered into it. In effect the tool is as only as good as the fool that uses it, so "a fool with a tool is still a fool"–a common criticism from industry [12]. While a business (or other) modeller may be no fool, there is no guarantee that a model created using CGs is correct in terms of its validity. In their exploration of the given problem using CGs, the modeller may have a misconception of the system being modelled or may simply make mistakes in its construction–things that still conform to the syntax and grammar but result in an invalid model. The current situation is too complicated, and presents an unwarranted burden on the modeller.

Formal Concept Analysis (FCA) claims to add mathematical rigour to the logical rigour captured in CGs [10]. *CGFCA* reveals FCA's effectiveness in this respect, thereby moving away from the current situation with its unnecessary complications as described above. We now further test this effectiveness using business modelling as the more comprehensive illustration.

5.2 Understanding the Complications

The CG Fig. 43 describes the components of a software application module that is part of an information system in an organisation. Applying *CGFCA* as above, Fig. 44 is the corresponding FCL for this CG Figure. The human business modeller draws this CG to capture the entities as CG Concepts and the CG relations between them. The detailed meaning of each entity and relation is discussed elsewhere [15], but for the purposes of our understanding the application module is denoted by the CG Concept: [Application Module: am1A]. The referent 'am1A' uniquely identifies the application module. The remaining CG Concepts and relations flow down from [Application Module: am1A]. This is validated by Application Module: am1A being at the supremum (topmost) Formal Concept of the lattice, Fig. 44. The modeller requires each referent throughout this CG to be unique and have no cycles in it. Figure 44 evidences that the Application Module as a CG is accurately captured. In practice this is unlikely to be the case. What are the complications in drawing the model that can undermine its validity, and how are these complications revealed by *CGFCA* and the lattice?

Arrow Direction. A common mistake or misconception that a modeller can make is to draw the arrows the wrong way round. This is a complication that may seem obvious on close inspection of the CG but nonetheless easily occurs even in introductory CGs despite proof-checking [14]. The syntax of the CG is correct–i.e. it is still a digraph (directed graph)–but this act results a semantic error. Figure 46, which is an extract of the lattice for the CG Fig. 45 shows that Application Module: am1A is not at the supremum; its place is taken by Organizational Unit: ou1A. This change of input is also shown by the CGFCA report: Inputs:"Organizational Unit: ou1A". The modeller is alerted to this deficiency because the arrows between [Application Module: am1A] and

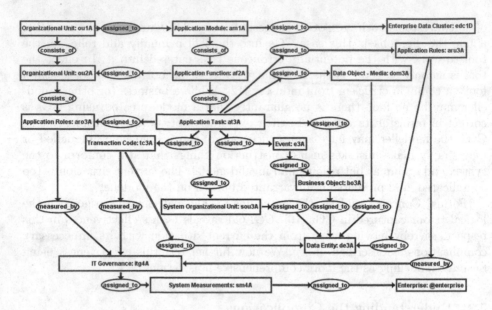

Fig. 45. Application module CG, wrong arrow direction

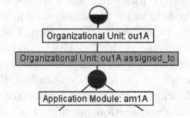

Fig. 46. Application module FCL extract, wrong arrow direction

[Organizational Unit: ou1A] in CG Fig. 45 were accidentally drawn the other way, unlike in the correctly-drawn previous CG Fig. 43. The modeller can then correct the model. The modeller may want to manually record the mistake for future reference, by shading the 'offending' (assigned_to) CG relation as shown in Fig. 45.

Mispointed Arrows. The CG Fig. 47 highlights another common mistake (or misconception) where a CG relation is pointed to the wrong CG Concept. In this case it's [Transaction Code: tc3A] → (assigned_to) → [IT Governance: itg4A]. It should be [Transaction Code: tc3A] → (assigned_to) → [Data Entity: de3A]. For convenience the offending relation is highlighted in Fig. 47. In practice the modeller would run *CGFCA* then generate the FCL (of which Fig. 48 is an extract) before highlighting the incorrect CG. From FCL Fig. 47 the modeller notices that assigned_to;measured_by is incorrectly concatenated in Transaction Code: tc3A assigned_to;measured_by.

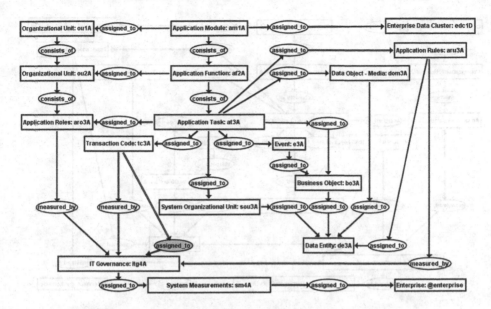

Fig. 47. Application module CG extract, assigned_to mispointed

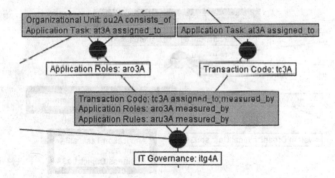

Fig. 48. Application module FCL extract, assigned_to mispointed

Such concatenations were demonstrated earlier by the FCL Fig. 20 (i.e. `Cat:` `Bumbles sleeps-on;sits-on;prefers`), which was correct in CG Fig. 19 but incorrect in CG Fig. 47. Again a close inspection of CG Fig. 47 would reveal this complication, but it can easily happen in practice.

Unwanted Cycles. While cycles may be deliberate, in many cases including this business modelling scenario they point to a mistake or misconception. That is the case of the CG Fig. 49 that emerges in the FCL of which Fig. 50 is an extract. The cycle is still rather subtle however as the FCA attributes `Business` `Object: bo3A assigned_to`, `Event: e3A assigned_to` and `Application Task:` `at3A assigned_to` are spread across two Formal Concepts before the FCA objects

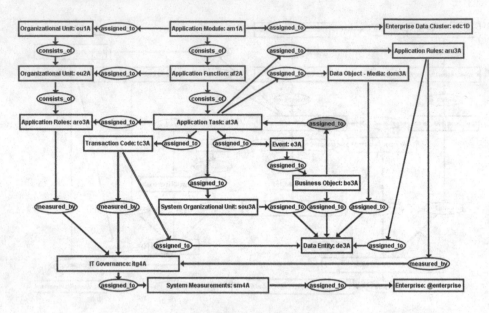

Fig. 49. Application module CG, cycle

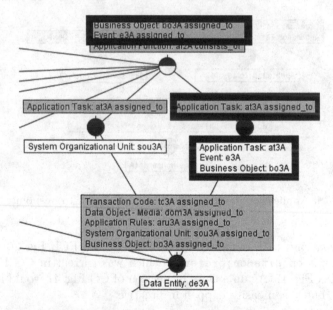

Fig. 50. Application module FCL extract, cycle

Business Object: bo3A, Event: e3A assigned_to and Application Task: at3A
are reached showing that the CG Concepts in the attributes (eventually) point to
themselves as the CG Concept denoted by the FCA object. The *CGFCA* report
brings it most easily to light:

Cycle: Application Task: at3A - assigned_to - Event: e3A -
assigned_to - Business Object: bo3A - assigned_to - Application
Task: at3A.

The FCL is nonetheless of value as the Formal Concept that has the three
attributes listed above (i.e. Business Object: bo3A assigned_to, Event:
e3A assigned_to and Application Task: at3A assigned_to) doesn't have
its own object (i.e. target CG Concept). This is highlighted by the bottom half
of this Formal Concept's circle being transparent due to the other dependencies
in the FCL. The modeller sets those other dependencies aside, as (s)he has identified that the unwanted cycle is the issue and its correction may also resolve any
other suspect dependencies (which it does). The cause? That common error of
a relation with the arrows pointing the wrong way i.e. the (assigned_to) CG
relation that points to [Application Task: at3A] from [Business Object:
bo3A] when the CG relation should be the other way round. This time it causes
a cycle as a revisit to the CG Fig. 49 and following this CG relation–using the
CGFCA report as our guide–brings the cycle to light. For the record, the offending (assigned_to) is shaded in the CG Fig. 49. The cycle in the FCL Fig. 50 is
also highlighted by the rectangles with thick black borders.

Invalid CG Referents. Remember in Fig. 22 i.e. Cat;Mat;Pet: Gwyn, Gwyn
became not only a Pet Cat but a Mat too! This common error appears in the
CG Fig. 51 and becomes evident in FCL Fig. 52, where [Application Roles:
ar3A] and [Application Rules: ar3A] accidentally share the same referent
(ar3A); an easy error to make especially as the CG Type Labels Application

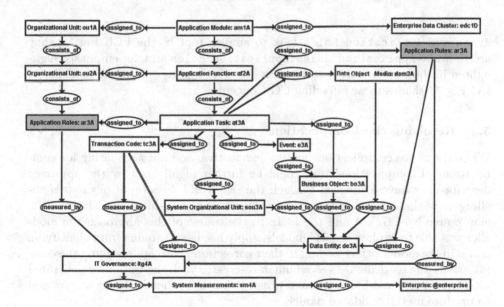

Fig. 51. Application module CG, ar3A referent

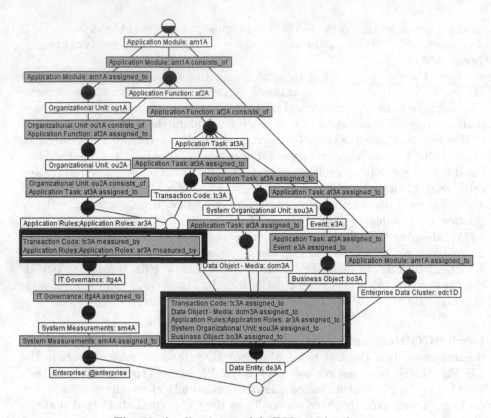

Fig. 52. Application module FCL, ar3A referent

Rules and Application Roles look so similar too! In the FCL Fig. 52 they are shown as Application Rules;Application Roles and, for emphasis, highlighted in thick black border rectangles. Likewise, and as before, the modeller in CG Fig. 51 shades these offending CG Concepts.

5.3   Resolving the Complications

While the above complications are not exhaustive, and not accounting for combinations of complications that could be further highlighted by the approach described, we have evidenced through the real-world scenario of business modelling how the human modeller as a practitioner (business or otherwise) is empowered by *CGFCA* and the FCL. In the course of this approach the modeller was able to explore the CG models, apply his/her reasoning from identifying issues in the models, thus leading to their correction. Through resolving the complications, the modeller acts as a human co-creator with the computer-generated *CGFCA* reports and FCLs (Formal Concept Lattices) thereby being empowered to produce useful, validated models.

6 Related Work

CGFCA originated with a comparative study to Wille's Concept Graphs as stated earlier, revealing the comparative advantages of *CGFCA* [5,20]. *CGFCA*–hence *Triples2Binaries*–is however now at a level of maturity that it can play a useful role whilst recognising the existence of other FCA approaches to triple-based structures, such as Relational Concept Analysis (RCA), EL-Implications and Graph-FCA [6,7,16]. Extensive comparative studies in this arena already exist, pre-*CGFCA* [13]. While *CGFCA* fulfills the scope of our study, there is value in an up-to-date comparative study that includes *CGFCA*. Such work may help to identify how all the approaches may best work together for directed graphs and FCA.

7 Concluding Remarks and Further Work

As well as providing the capability to explore, reason with and validate directed graphs (digraphs), the FCL representation of CGs are arguably more readable. As we have seen, the arcs (the arrows) in a CG can lead in any direction. In a large, complex CG it can be difficult to trace and compare pathways through it, even more so where there are co-referent links. All FCL pathways are aligned in a top-to-bottom (inputs to outputs), hierarchical manner and co-referents can be automatically joined to make more apparent their connections and place in the graph.

Future work will continue to develop representative exemplars, a worthwhile endeavour given the value demonstrated by this paper. Furthermore since we have set the context as exploring and validating digraphs through triples to binaries rather than just CGs, the further work intends to include directed triples modelled by practitioners in UML, RDF, OWL, the Entity-Relation Diagram and linked data as alluded to earlier.

Meanwhile we have demonstrated that *CGFCA*–hence *Triples2Binaries*–presents a formal method that exploits CGs as digraphs through the application of Formal Concept Analysis (FCA). FCA elucidates key features of CGs such as pathways and dependencies, inputs and outputs, cycles, and joins. Given the prevalence of digraphs, the practitioner is consequently empowered in exploring, reasoning with and validating their models in understanding real-world phenomena.

References

1. Charger - a conceptual graph editor. http://charger.sourceforge.net/. Accessed 02 Jan 2018
2. Cogui. http://www.lirmm.fr/cogui/. Accessed 02 Jan 2018
3. Andrews, S.: In-Close2, a high performance formal concept miner. In: Andrews, S., Polovina, S., Hill, R., Akhgar, B. (eds.) ICCS 2011. LNCS (LNAI), vol. 6828, pp. 50–62. Springer, Heidelberg (2011). https://doi.org/10.1007/978-3-642-22688-5_4

4. Andrews, S., Hirsch, L.: A tool for creating and visualising formal concept trees. In: CEUR Workshop Proceedings, vol. 1637, pp. 1–9 (2016)
5. Andrews, S., Polovina, S.: A mapping from conceptual graphs to formal concept analysis. In: Andrews, S., Polovina, S., Hill, R., Akhgar, B. (eds.) ICCS 2011. LNCS (LNAI), vol. 6828, pp. 63–76. Springer, Heidelberg (2011). https://doi.org/10.1007/978-3-642-22688-5_5
6. Baader, F., Distel, F.: A finite basis for the set of $\mathcal{EL}$-implications holding in a finite model. In: Medina, R., Obiedkov, S. (eds.) ICFCA 2008. LNCS (LNAI), vol. 4933, pp. 46–61. Springer, Heidelberg (2008). https://doi.org/10.1007/978-3-540-78137-0_4
7. Ferré, S., Cellier, P.: Graph-FCA in practice. In: Haemmerlé, O., Stapleton, G., Faron Zucker, C. (eds.) ICCS 2016. LNCS (LNAI), vol. 9717, pp. 107–121. Springer, Cham (2016). https://doi.org/10.1007/978-3-319-40985-6_9
8. Ganter, B., Wille, R.: Formal Concept Analysis: Mathematical Foundations. Springer, Heidelberg (2012). https://doi.org/10.1007/978-3-642-59830-2
9. Harary, F.: Structural Models: An Introduction to the Theory of Directed Graphs. Wiley, New York (1965)
10. Hitzler, P., Scharfe, H.: Conceptual Structures in Practice. CRC Press, Boca Raton (2009)
11. Koehler, K.R.: Directed graphs (2012). http://kias.dyndns.org/comath/33.html
12. Parker, L., HP OpenView Business Unit: A fool with a tool is still a fool! HP Open View (2001)
13. Poelmans, J., Ignatov, D.I., Kuznetsov, S.O., Dedene, G.: Review: formal concept analysis in knowledge processing: a survey on applications. Expert Syst. Appl. **40**(16), 6538–6560 (2013)
14. Polovina, S.: An introduction to conceptual graphs. In: Priss, U., Polovina, S., Hill, R. (eds.) ICCS-ConceptStruct 2007. LNCS (LNAI), vol. 4604, pp. 1–14. Springer, Heidelberg (2007). https://doi.org/10.1007/978-3-540-73681-3_1
15. Polovina, S., Scheruhn, H.-J., von Rosing, M.: Modularising the complex meta-models in enterprise systems using conceptual structures. In: Developments and Trends in Intelligent Technologies and Smart Systems, pp. 261–283. IGI Global, Hershey (2018). ID: 189437
16. Rouane-Hacene, M., Huchard, M., Napoli, A., Valtchev, P.: Relational concept analysis: mining concept lattices from multi-relational data. Ann. Math. Artif. Intell. **67**(1), 81–108 (2013)
17. Sowa, J.F.: Conceptual Structures: Information Processing in Mind and Machine. Addison-Wesley Publishing, Reading (1983)
18. Sowa, J.F.: Conceptual graph examples. http://www.jfsowa.com/cg/cgexampw.htm
19. Sowa, J.F.: Conceptual graphs. In: Handbook of Knowledge Representation, Foundations of Artificial Intelligence, vol. 3, pp. 213–237. Elsevier, Amsterdam (2008)
20. Wille, R.: Conceptual graphs and formal concept analysis. In: Lukose, D., Delugach, H., Keeler, M., Searle, L., Sowa, J. (eds.) ICCS-ConceptStruct 1997. LNCS, vol. 1257, pp. 290–303. Springer, Heidelberg (1997). https://doi.org/10.1007/BFb0027878

Subjective Bayesian Networks and Human-in-the-Loop Situational Understanding

Dave Braines[1,2], Anna Thomas[1], Lance Kaplan[3], Murat Şensoy[2,6],
Jonathan Z. Bakdash[4,5], Magdalena Ivanovska[7], Alun Preece[2],
and Federico Cerutti[2(✉)]

[1] IBM Hursley Park, Winchester, UK
[2] Cardiff University, Cardiff, UK
CeruttiF@cardiff.ac.uk
[3] U.S. Army Research Laboratory, Adelphi, USA
[4] U.S. Army Research Laboratory South Field Element,
The University of Texas, Dallas, USA
[5] Texas A&M Commerce, Commerce, USA
[6] Ozyegin University, Istanbul, Turkey
[7] University of Oslo, Oslo, Norway

Abstract. In this paper we present a methodology to exploit human-machine coalitions for situational understanding. Situational understanding refers to the ability to relate relevant information and form logical conclusions, as well as identify gaps in information. This process for comprehension of the meaning information requires the ability to reason inductively, for which we will exploit the machines' ability to 'learn' from data. However, important phenomena are often rare in occurrence with high degrees of uncertainty, thus severely limiting the availability of instance data for training, and hence the applicability of many machine learning approaches. Therefore, we present the benefits of Subjective Bayesian Networks—i.e., Bayesian Networks with imprecise probabilities—for situational understanding, and the role of conversational interfaces for supporting decision makers in the evolution of situational understanding.

1 Introduction

Human situational understanding is filled with inductive reasoning. Say you just landed at Heathrow Airport in London, UK: the sun is blazing in the sky and

This research was sponsored by the U.S. Army Research Laboratory and the U.K. Ministry of Defence under Agreement Number W911NF-16-3-0001. The views and conclusions contained in this document are those of the authors and should not be interpreted as representing the official policies, either expressed or implied, of the U.S. Army Research Laboratory, the U.S. Government, the U.K. Ministry of Defence or the U.K. Government. The U.S. and U.K. Governments are authorized to reproduce and distribute reprints for Government purposes notwithstanding any copyright notation hereon.

© Springer International Publishing AG, part of Springer Nature 2018
M. Croitoru et al. (Eds.): GKR 2017, LNAI 10775, pp. 29–53, 2018.
https://doi.org/10.1007/978-3-319-78102-0_2

a glorious warm temperature of 23 Celsius (74 Fahrenheit) welcomes you in the South of Britain. On the basis of this observation, it is rational to conclude that usually the South of Britain enjoys lovely weather, especially if the same happens the second day, the third day, and the fourth day of your visit. From a human perspective, general rules and thus understanding are, therefore, often derived on the basis of scarce data/information. Consequently, human decision making frequently exhibits heuristics and biases rather than following rationality [11].

Sometimes limited data is not a problem, especially in those cases where we can have access to an *oracle*, mostly an expert in the domain. You might receive a useful piece of information from a friend who lived in the South of Britain for years, or you can access historical data and statistics showing that the South of Britain does not usually enjoy lovely weather, and therefore this apparent normality is in fact an exception. Oracles can help in overcoming scarcity of actual data through access to other information or rules that are relevant to the domain.

As humans we, therefore, apply analyses and judgements to relevant information "to determine the relationships of the factors present and form logical conclusions concerning threats, opportunities, and gaps in information" [7]. This is *situational understanding*.

Machine learning approaches are potentially powerful allies in situational understanding [3]. This is because machine learning algorithms are able to efficiently handle large quantities of information, which is extremely useful to support inductive reasoning in situational understanding, as well as deriving logical conclusions. However, they are generally useless for identifying gaps in information as well as in providing insights, such as those that could be provided by oracles. Moreover, many of the best algorithms for machine learning often assume the existence of a large training set with independent and identically distributed (IID) data. Algorithms for data with limited instances (class imbalanced) is a specific research area [5], as are algorithms for non-IID data [24]. Unfortunately, the assumption of large amounts of balanced and IID data tends to unrealistic in the real-world. Rare events often arise from multiple *dependent* factors: for example, the risk of political instability is a combination of corruption, illicit activities, and organised crime [8].

The need for less training data and modeling of underlying dependencies is particularly important in situational understanding problems where many important phenomena will be rare in occurrence, severely limiting the availability of instance data and, hence, the applicability of many machine learning approaches, including Bayesian and Deep Learning [16] approaches. Coupled with this, supporting human analysts in terms of more effective communication of uncertain information is also a key issue in situational understanding problems [6].

In this paper—that is an extended version of [2]—we propose a human-machine coalition partnership for real-world situational understanding by exploiting the strengths of each member in the coalition. Machines' strengths are linked to data analysis, and we explicitly address the unrealistic assumption of large training sets that could undermine the role of machine agents in such a human-machine coalition. Moreover, human experts are usually considered useful oracles, and we need to provide useful human-machine interfaces in order to support co-design and co-evolution of the coalition for situational understanding. Specifically, we consider a system within which the human agents can contribute to or correct the machine agent parts of the system.

To exemplify our proposal, we discuss a running example about the German stock market in Sect. 2, and in Sect. 3 we exploit one of the machines' strengths: performing inductive reasoning with quantitative measures such as probabilities. We discuss a robust approach to handling uncertain information from a rather scarce dataset, namely Subjective Bayesian Networks, an extension of Bayesian Networks using uncertain probabilities. This helps us towards overcoming one of the main issues related to Bayesian networks: the lack of information about the certainty of the trained model.

We then show, in Sect. 4, that Subjective Bayesian Networks are well suited for situational understanding. Our tests show that they provide more accurate results compared to other approaches to Bayesian networks with uncertain probabilities, such as Credal networks [23] and belief networks [20].

Finally, in Sect. 5 we summarise an evaluation we performed with a focus group on the usage of conversational interfaces for co-designing a Subjective Bayesian Network and using it for situational understanding.

2 Human-Machine Coalitions for Situational Understanding

Let us suppose you are an advisor for investors who want to enter the German stock market. For brevity, let us suppose that a colleague has provided the two high-level dependency networks depicted in Fig. 1, showing on the one hand dependencies between Daimler, BMW, Continental, Porsche, and Volkswagen (automotive companies); and on the other hand dependencies between Bayer, Henkel, and Beiersdorf (cosmetic companies). These dependencies suggest that the stock prices of those companies are linked such that a significant variation of the stock price of Daimler will influence a variation in the stock price of BMW.

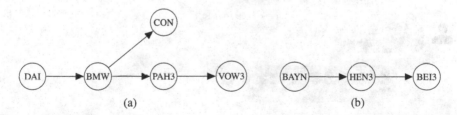

Fig. 1. German automotive (a) and cosmetic (b) company dependency networks provided as input

Table 1. Companies considered from the German stock market in Figs. 1 and 2

	Company	Comment
BAYN	Bayer	Pharmaceutical company
BEI3	Beiersdorf	Cosmetic company
BMW	BMW	Automotive manufacturer
CON	Continental	Tyre manufacturer
DAI	Daimler	Automotive manufacturer
HEN3	Henkel	Cosmetic company
PAH3	Porsche	Automotive manufacturer
VOW3	Volkswagen	Automotive manufacturer

Let us suppose you have the privilege of using our conversational interface for interacting with such dependencies networks, see Fig. 2. Among other activities, such as explaining the dependencies and exploring what-if scenarios such a conversational interface would allow you also to express additional information, in particular that there is a dependency between Bayer and Daimler thus *de facto* providing a machine with domain knowledge unavailable before. This enables the human user to, therefore, act as an *oracle*, contributing relevant information to the machine agent based on their wider knowledge of the domain in question.

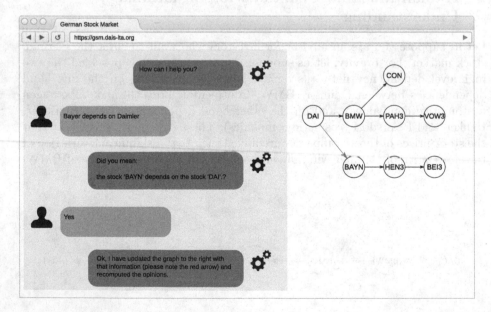

Fig. 2. Mockup depicting the action of updating a dependency network through our proposed conversational interface. Other speech acts envisaged for such an interface include "Explain dependencies..." and "What happens if ..."

Indeed, Daimler and Bayer are regularly traded by over-the-counter (OTC) list shares[1] such as INTL FCStone Financial.[2]

3   Reasoning Under Uncertainty with Limited Data

3.1   Dealing with Uncertainty: Subjective Logic

Subjective logic is a formalism for reasoning under uncertain probabilistic information [10]. It expands the notion of a probability value to a distribution of possible probabilities. This paper considers binary variables such as X that can take on the value of true or false, i.e., $X = \mathfrak{x}$ or $X = \bar{\mathfrak{x}}$. The value of X does change over different instantiations, and there is an underlying ground truth value for the probability $p_X(x)$ of taking on the value in the domain $\mathbb{X} = \{\mathfrak{x}, \bar{\mathfrak{x}}\}$. In general, the variable can take on one of K mutually exclusive values.

A subjective opinion can be formed by directly observing N_{ins} independent instantiations of X. If over these instantiations, n_x times $X = \mathfrak{x}$, $n_{\bar{x}} = N_{ins} - n_x$ times $X = \bar{\mathfrak{x}}$ and assuming an uninformative uniform prior, then the posterior knowledge of the ground truth outcome probability of X is known to follow the beta distribution

$$f_\beta(p_x|\omega_X) = \frac{1}{\beta(\alpha_x, \alpha_{\bar{x}})} p_x^{\alpha_x - 1}(1 - p_x)^{\alpha_{\bar{x}} - 1} \tag{1}$$

for $0 \le p_x \le 1$, where $\beta(\cdot)$ is the beta function and the beta parameters $\alpha = [\alpha_x, \alpha_{\bar{x}}] = [n_x + 1, n_{\bar{x}} + 1]$ are one particular representation of the opinion ω_X. The opinion ω_X in belief space is a tuple of belief $b_X = \frac{n_x}{s_X}$, disbelief $d_X = \frac{n_{\bar{x}}}{s_X}$ and uncertainty $u_X = \frac{2}{s_X}$, where $s_X = \alpha_x + \alpha_{\bar{x}}$ is the Dirichlet strength. Therefore, a tuple $\langle b_X, d_X, u_X \rangle$ identifies a point in a 3D space. However, since the belief masses are positive and sum up to one, such a 3D space can be flattened into a 2D triangle, as depicted in Fig. 3. Following [10, p. 49] we can partition the 2D space of subjective logic opinions for (lossy) representation using fuzzy natural language terms such as "High Confidence" and "Very Likely". Such terms can be made even more consumable for human users when embedded within larger natural language sentences such as: "When BAYN stock price changes, there is *high confidence* that HEN3 stock price is *very likely to change*" that can summarise the subjective opinion $\langle 0.8, 0.1, 0.1 \rangle$.

In this paper, it will be convenient to represent the subjective opinion ω_X by the mean and Dirichlet strength of the corresponding beta distribution. The mean represents the projected probability that converts the opinion into the pignistic probabilities, and is given by

$$P_X(\mathfrak{x}) = \frac{\alpha_x}{s_X} \quad \text{and} \quad P_X(\bar{\mathfrak{x}}) = \frac{\alpha_{\bar{x}}}{s_X}. \tag{2}$$

---

[1] OTC trades refers to stock trades via a dealer network.
[2] https://goo.gl/1Truuv (on 4th May 2017).

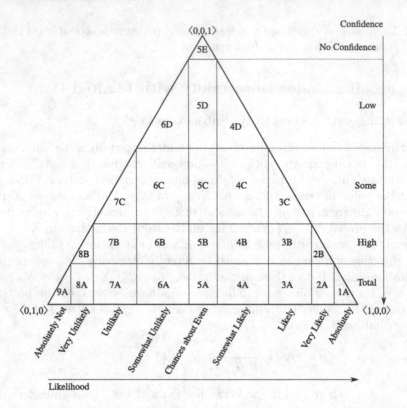

Fig. 3. Subjective logic 2D triangle and areas for fuzzy labels, adapted from [10, p. 49].

The variance of the corresponding beta distribution,

$$\sigma_X^2 = \frac{P_X(\mathfrak{x})P_X(\bar{\mathfrak{x}})}{s_X + 1}, \tag{3}$$

is a function of the projected probabilities and Dirichlet strength of the subjective opinion. This expression is used in the experiments to predict the root mean squared error between the projected probability $P_X(x)$ and the actual ground truth $\rho_X(x)$. Subjective opinions naturally extend to subjective conditional opinions, where for example, the opinion for X conditioned on Y and Z is interpreted as the set $\{\omega_{X|y,z} : y \in \mathbb{Y}, z \in \mathbb{Z}\}$, and $\omega_{X|y,z}$ represents the effective number of times that $X = \mathfrak{x}$ or $X = \bar{\mathfrak{x}}$ when $Y = y$ and $Z = z$ while jointly observing X, Y, and Z.

3.2 Dealing with Limited Data: Subjective Bayesian Network

The Subjective Bayesian network (SBN) was first proposed in [9], and it is an uncertain Bayesian network where the conditionals are subjective opinions instead of dogmatic probabilities. In other words, the conditional probabilities

are known within a beta distribution. A SBN reflects the knowledge about a
Bayesian network when limited historical data is used to learn the conditionals.
The inference in SBN leads to an opinion about the marginal probability of all
the unobserved variables conditioned on the values of the observed variables.
While different types of SBNs were discussed in [9], this paper focuses on the
type that uses the beta distribution interpretation of the subjective opinion to
compute uncertainty. This section reviews subjective belief propagation (SBP),
which was introduced for trees in [12] and extended for singly-connected networks
in [13] for this class of SBNs.

SBP extends the Belief Propagation (BP) inference method of Pearl [19]. In
BP, π- and λ-messages are passed from parents and children, respectively, to
a node, i.e., variable. The node uses these messages to formulate the inferred
marginal probability of the corresponding variable. The node also uses these
messages to determine the π- and λ-messages to send to its children and parents,
respectively. In SBP, the π- and λ-messages are subjective opinions characterized
by a projected probability and Dirichlet strength.

The SBP formulation approximates output messages as beta-distributed random variables using the methods of moments and a first-order Taylor series
approximation to determine the mean and variance of the output messages in
light of the beta-distributed input messages. The details of the derivations are
provided in [12,13]. Given a node X with m parents U_i for $i = 1, \ldots, m$, the
subjective opinions of the π-messages sent to X are characterized by the projected probabilities $\pi_{U_i,X}(x)$ and Dirichlet strengths $s_{\pi_{U_i,x}}$. Likewise, given that
X has k children Y_j for $j = 1, \ldots, k$, the subjective opinions of the λ-messages
sent to X are characterised by the projected probabilities $\lambda_{U_i,X}(x)$ and Dirichlet
strengths $s_{\lambda_{U_i,X}}$. Node X processes these opinions to form the fused π opinion

$$\pi_X(x) = \sum_{u_1,\ldots,u_m} P(x|u_1,\ldots,u_m) \prod_{i=1}^{m} \pi_{U_i,X}(u_i), \tag{4}$$

$$s_{\pi_X} = \frac{\pi_X(x)(1 - \pi_X(x))}{\sigma^2_{\pi_X}} - 1, \tag{5}$$

where the variance $\sigma^2_{\pi_X} = V_{\pi_X} - \pi^2_X(x)$,

$$V_{\pi_X} = \sum_{u_1,\ldots,u_m} \sum_{u'_1,\ldots,u'_m} g(x,x;u_1,\ldots,u_m;u'_1,\ldots,u'_m) \cdot \prod_{i=1}^{m} h(u_i,u'_i), \tag{6}$$

$$g(x,x';u_1,\ldots,u_m;u'_1,\ldots,u'_m) = p_{x|u_1\ldots u_m} p_{x|u'_1\ldots u'_m}$$
$$+ (-1)^{x \neq x'} \delta_{\mathbf{u},\mathbf{u'}} \frac{p_{x|u_1\ldots u_m}(1 - p_{x|u_1\ldots u_m})}{s_{X|u_1\ldots u_m} + 1}, \tag{7}$$

where $\mathbf{u}$ is an arbitrary joint assignment of the variables $U_1, \ldots, U_m$,

$$\delta_{\mathbf{u},\mathbf{u'}} = \begin{cases} 1, & \text{if } u_j = u'_j, \text{ for } j = 1,\ldots,m \\ 0, & \text{otherwise} \end{cases}$$

is the Kronecker delta function, and

$$h_\pi(u_i, u_i') = \pi_{U_i,X}(u_i)\pi_{U_i,X}(u_i') + (-1)^{u_i \neq u_i'}\frac{\pi_{U_i,X}(u_i)(1 - \pi_{U_i,X}(u_i))}{s_{\pi_{U_i,X}} + 1}. \tag{8}$$

The fused λ-message is

$$\lambda_X(x) = \alpha_\lambda \prod_{j=1}^{k} \lambda_{Y_j,X}(x), \tag{9}$$

$$s_{\lambda_X} = \left(\sum_{j=1}^{k} \frac{\lambda_X(\mathfrak{x})\lambda_X(\bar{\mathfrak{x}})}{\lambda_{Y_j,X}(\mathfrak{x})\lambda_{Y_j,X}(\bar{\mathfrak{x}})}\frac{1}{s_{\lambda_{Y_j,X}} + 1}\right)^{-1} - 1,$$

where α_λ is a normalizing constant so that $\lambda_X(x)$ sums to one over its domain $\mathbb{X}$.

The π and λ-opinions are fused to determine the marginal opinion for node X:

$$P_X(x|o) = \alpha_f \pi_X(x)\lambda_X(x), \tag{10}$$

$$s_X = \left(\frac{P_X(\mathfrak{x})P_X(\bar{\mathfrak{x}})}{\pi_X(\mathfrak{x})\pi_X(\bar{\mathfrak{x}})}\frac{1}{s_{\pi_X} + 1} + \frac{P_X(\mathfrak{x})P_X(\bar{\mathfrak{x}})}{\lambda_X(\mathfrak{x})\lambda_X(\bar{\mathfrak{x}})}\frac{1}{s_{\lambda_X} + 1}\right)^{-1} - 1,$$

where α_f is also a normalizing constant.

The opinion for the message that node X sends to parent U_i is

$$\lambda_{X,U_i}(u_i) = \alpha_b \sum_x \lambda_X(x) \sum_{\{u_1,\ldots,u_m\}\setminus\{u_i\}} P(x|u_1,\ldots,u_i,\ldots,u_m) \cdot \prod_{j\neq i} \pi_{U_j,X}(u_j), \tag{11}$$

$$s_{\lambda_{X,U_i}} = \frac{\lambda_{X,U_i}(u_i)(1 - \lambda_{X,U_i}(u_i))}{\sigma^2_{\lambda_{X,U_i}}} - 1, \tag{12}$$

where

$$\sigma^2_{\lambda_{X,U_i}} = \alpha_b^2\left(\lambda^2_{X,U_i}(\bar{x})\sigma^2_{uu} + \lambda^2_{X,U_i}(x)\sigma^2_{\bar{u}\bar{u}} - 2\lambda_{X,U_i}(x)\lambda_{X,U_i}(\bar{x})\sigma^2_{u\bar{u}}\right), \tag{13}$$

$$\sigma^2_{zv} = \sum_x \sum_{x'} h_\lambda(x,x') \sum_{\{u_1,\ldots,u_m\}\setminus\{z\}} \sum_{\{u_1',\ldots u_m'\}\setminus\{v\}}$$
$$g(x,x';u_1,\ldots,z,\ldots,u_m;u_1',\ldots,v,\ldots,u_m')\prod_{j\neq i} h_\pi(u_j,u_j'), \tag{14}$$

and

$$h_\lambda(x,x') = \lambda_X(x)\lambda_X(x') + (-1)^{x \neq x'}\frac{\lambda_X(x)(1 - \lambda_X(x))}{s_{\lambda_X} + 1}, \tag{15}$$

and α_b is a normalizing constant.

Finally, the opinion message sent to the children of X are

$$\pi_{X,Y_j}(x) = \alpha_\pi \prod_{i \neq j} \lambda_{Y_i,X}(x)\pi_X(x), \tag{16}$$

$$s_{\pi_{X,Y_j}} = \left(\frac{\pi_{X,Y_j}(\mathfrak{x})\pi_{X,Y_j}(\bar{\mathfrak{x}})}{\pi_X(\mathfrak{x})\pi_x(\bar{\mathfrak{x}})} \frac{1}{s_{\pi_X}+1} + \sum_{i \neq j} \frac{\pi_{X,Y_j}(\mathfrak{x})\pi_{X,Y_j}(\bar{\mathfrak{x}})}{\lambda_{Y_i,X}(\mathfrak{x})\lambda_{Y_i,X}(\bar{\mathfrak{x}})} \frac{1}{s_{\lambda_{Y_i,X}}+1} \right)^{-1} - 1,$$

where α_π is a normalizing constant.

The equations for the projected probability updates in SBP mirror the updated equations in standard belief propagation due to the first-order Taylor approximation. Actually, the normalizing constants α_λ and α_β are superfluous in standard belief propagation, but necessary in SBP so that the λ message are proper subjective opinions. In short, SBP provides the same answer as belief propagation in the mean value. The difference is that SBP also provides a quantification of the uncertainty through the Dirichlet strength. On a technical note, SBP will actually increase the Dirichlet strength as computed in the update equations to ensure that all belief values are non-negative. We refer the interested reader to [12,13] for more details. Finally, the information flow in SBP is exactly the same as in belief propagation. For the sake of comparison, a node can send a message to one particular neighbor once it receives messages from all of its other neighbors.

4   Experimentation

4.1   Methodology

SBNs can learn a model of the domain with a very limited number of observations; however, the inferred opinions through such a network will become more certain as the number of observations increases. To measure how well these models can be learned with limited data and measure the uncertainty associated with the inferences, we build gold standard models, which are Bayesian networks that are generated using a much larger number of observations. The gold standard models are Bayesian networks with completely certain conditional probabilities that we treat as the ground truth.

For structure learning of the gold standard models, we used the well-known K2 algorithm [17]. The K2 algorithm is used to learn the best structure of a singly connected Bayesian network to represent the interactions between the random variables. The resulting network serves as a surrogate for a subject matter expert who would use their background knowledge to create the network structure, for example, via the conversational interface (see Fig. 2). Further discussion on this topic is provided in the conclusion of the paper. Then, the conditional and marginal probabilities at each node of the network are calculated in the traditional manner using the entire available data.

We use real data to evaluate the quality of the uncertainty (or Dirichlet strength) in the subjective opinions inferred by SBP to represent the actual spread between the corresponding 'projected' and 'ground truth' probabilities that are well captured by the gold standard models. The full data is then divided into non-overlapping segments of N_{ins} instantiations (i.e., observations). Each segment represents the sparse data that would actually be available to train a SBN. A SBN is trained for each segment, and the set of exterior nodes, i.e., nodes with one single neighbour (either a parent or child), are considered to be observed. For each combination of possible values for these exterior nodes, the marginal opinions for the interior nodes are inferred by SBP. Likewise, to establish the ground truth, the marginal probabilities are inferred by standard belief propagation using the underlying gold standard Bayesian network for the same values of the observed exterior nodes. Then, the marginal opinions and ground truths for all interior nodes are determined over all combinations of observed values and non-overlapping segments. Finally, the uncertainty of the marginal opinions is evaluated.

To evaluate the quality of the derived uncertainty, the actual root mean squared error (RMSE) between the projected and ground truth probabilities is calculated. Next, the predicted RMSE is computed without knowledge of the ground truth, as the square root of the average variance predicted from the opinions via (3). The similarity between the actual and predicted RMSE is one way to establish the quality of the uncertainty in the subjective opinions that are to characterise the spread between the projected and actual probabilities.

An even more precise method to determine the quality of the uncertainty characterisation is to establish γ-confidence intervals from the opinions to capture the fraction of γ ground truths within these intervals. One then tabulates the fraction of times that the actual ground truth falls within the confidence interval. This is done for various values of $\gamma \in [0, 1]$, and the plot of the actual $\hat{\gamma}$ and the desired γ should follow a straight line as it should be the case that $\hat{\gamma} \approx \gamma$. A more detailed discussion can be found in [14]. The quality of the inferred subjective opinion ω_X should be judged on how well its expression of uncertainty captures the spread between its projected probability and the actual ground truth probability.

We compare the performance of SBP against previous methods for reasoning over uncertain probabilistic networks. Namely, we consider credal networks and belief networks, which are summarized below:

Credal Networks: A credal network over binary random variables extends a BN by replacing single probability values with closed intervals representing the possible range of probability values. The extension of Pearl's message-passing algorithm by the 2U algorithm for credal networks is described in [23]. This algorithm works by determining the maximum and minimum value (an interval) for each of the target probabilities based on the given input intervals. It turns out that these extreme values lie at the vertices of the polytope dictated by the extreme values of the input intervals. As a result, the computational complexity grows exponentially with respect to the number of parents nodes. For the sake of comparison, we assume that our subjective network elicited from the given data corresponds to a credal network in the following way: if $\omega_x = [b_x, b_{\bar{x}}, u_X]$ is a subjective opinion on the probability p_x, then we have $[b_x, b_x + u_X]$ as an interval corresponding to this probability in the credal network. It should be noted that this mapping from the Beta distribution to an interval is consistent with past studies of credal networks [15].

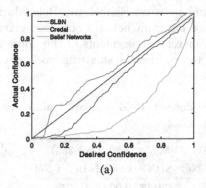 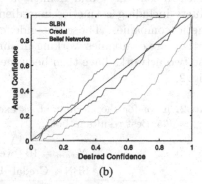

(a)   (b)

Fig. 4. Comparing SBN against Belief Networks and Credal with $N_{train} = 10$ (over 365) (a) and $N_{train} = 30$ (over 365) (b) for the German stock exchange data. Best closest to the diagonal.

Belief Networks: In [20], Smets introduced a computationally efficient method to reason over networks via Dempster-Shafer theory. It is an approximation of a valuation-based system. Namely, a (conditional) subjective opinion $\omega_X = [b_x, b_{\bar{x}}, u_X]$ from our SBN obtained from data is converted to the following belief mass assignment: $m(x) = b_x$, $m(\bar{x}) = b_{\bar{x}}$ and $m(x \cup \bar{x}) = u_X$. (Note that in the binary case, the belief function overlaps with the belief mass assignment). The method exploits the disjunctive rule of combination (DRC) to compose beliefs conditioned on the Cartesian product space of the binary power sets. This enables both forward propagation and backward propagation after inverting the belief conditionals via the generalized Bayes' theorem (GBT). By operating in the Cartesian product space of the binary power sets, the computational complexity grows exponentially with respect to the number of parents, similar to the 2U algorithm for credal sets and our SBP method.

4.2   German Stock Exchange Predictions

Let us consider the case where a machine learning system is used to mine data from the German Stock Market, Börse Frankfurt. To simplify the scenario, let us consider a binary variable per each company listed in Börse, where such a variable is *true* if there is a significant increase (i.e. +0.5%) in the company's stock value over a day, and *false* otherwise. Let us then suppose that a well-known off-the-shelf algorithm for structure learning of dependencies among selected variables, such as K2 [17], has been used. Using such an algorithm, the dependency networks highlighted in Figs. 1(a) and (b) are derived. Table 1 explains the variables used in the dependency networks.

Figure 1(a) shows how there is a dependency between Daimler stock varia-
tions and BMW; between BMW and Porsche; between Porsche and Volkswagen
(all automotive manufacturers); and between BMW and Continental, a tyre manufacturer. Similarly, Fig. 1(b) depicts the dependencies between Bayer—a pharmaceutical company—and Henkel—a company producing a variety of chemical
products including cosmetics ingredients; and between Henkel and Beiersdorf,
cosmetic companies. Those dependencies are far from being a surprise, given
that they are companies working in similar, or related, segments of the market.
These two networks have then been merged to produce the single network given
in Fig. 2.

Table 2. Error for the German stock exchange dataset. Gold standard trained with
$N_{train} = 365$. Best results in bold.

	$N_{train} = 10$ (over 365)			$N_{train} = 30$ (over 365)		
	SBN	Credal	Belief Net	SBN	Credal	Belief Net
Actual RMSE	**0.124**	0.198	0.176	**0.047**	0.062	0.075
Predicted RMSE	0.101	0.187	0.132	0.049	0.089	0.061

The gold standard Bayesian Network is obtained by using all available data
for (365 days) to determine the conditional probabilities. Then N_{train} days were
used to generate $floor(365/N_{train})$ SBNs. Binary values were generated for the
three nodes that have one edge, and the marginal probabilities (ground truth)
and marginal opinions were generated via belief propagation and subjective belief
propagation over the Bayesian and SBNs, respectively. Table 2 lists actual and
predicted RMSE for the different approaches using different amounts of observations. It indicates that SBN achieves pretty good error rate even with 10 days of
observations (sample size 2.74%) and the error decreases to 0.05 when 30 days of
data is used (sample size 8.21%). Figure 4 shows the ratio of the times the ground
truth falls within the bounds—set at various significance levels—when building
SBNs over 10 and 30 days. Our results indicate that SBN can capture the uncertainty more accurately than Credal networks and Belief Networks. Especially,
when $N_{train} = 30$, confidence level of the SBN is around the desired one, i.e.,
diagonal on the figures. Moreover, Table 2 lists actual and predicted RMSE for
our approach and the benchmark approaches when different amounts of observations are used. SBN is consistently able to predict an accurate RMSE.

4.3   Istanbul Stock Market Predictions

We also considered the dataset first derived in [1],[3] which considers stock
exchange returns for several indexes, including those listed in Table 3. It is quite
straightforward to derive a dependency network such as the one given in Fig. 6
between those indexes.

---

[3] https://goo.gl/XzAZUX (on 4th May 2017).

Standard & Poor's 500 index includes leading US companies and captures approximately 80% of available US market capitalisation. Those companies are trading heavily with the rest of the world, including Asia, and notably Japan; and with South America, notably Brazil. Moreover, Brazil's economy heavily affects the MSCI Emerging Markets Index. According to the Foreign Trade figures from the United States Census Bureau, within Europe, the US has a strong commercial partnership with Germany,[4] much stronger than with the second strongest commercial ally, namely the UK.[5] Therefore, it is straightforward to see how the return for Standard & Poor's has a significant statistical dependence with the German Stock Market. Moreover, with 15% of the imports coming from Germany, the UK economy is also significantly dependent on the German market[6] (instead Germany imports mostly from the Netherlands and exports mostly to the US).[7] Finally, the MSCI European Index return is heavily affected by Germany, the first economy in the European Union.

We also used this dataset of 536 entries to evaluate our approach using different amounts of observed data. Table 4 lists actual and predicted RMSE for our approach and the benchmark approaches when different amount of observations are used. It shows that SBN consistently predicts the error when trained either over 10 or 30 days, unlike the two other methods.

Figure 5 demonstrates our results in terms of γ-confidence intervals. Even for data of 10 days, the confidence for inferences with SBN only slightly diverges from the desired confidence levels. When training data is increased to 30 days, the confidence interval for SBN approximate the desired one very closely. Again, in this dataset, the best performance belongs to SDN in terms of γ-confidence intervals.

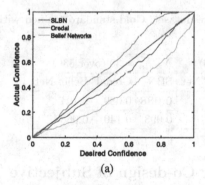

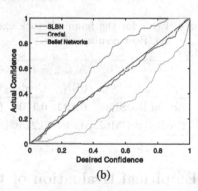

(a)   (b)

Fig. 5. Comparing SBN against Belief Networks and Credal with $N_{train} = 10$ (over 536) (a) and $N_{train} = 30$ (over 536) (b) for the Istanbul stock market data. Best closest to the diagonal.

---

[4] https://goo.gl/8PdBll (on 4th May 2017).
[5] https://goo.gl/n2V89z (on 4th May 2017).
[6] https://goo.gl/v1tXD4 (on 4th May 2017).
[7] https://goo.gl/ZPJLdR (on 4th May 2017).

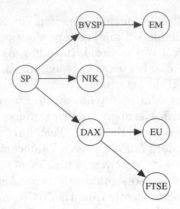

Fig. 6. Istanbul stock exchange data set [1] dependency network.

Table 3. Indexes considered from the Istanbul stock exchange data set [1] in Fig. 6.

	Comment
SP	Standard & Poor's 500 Index Return
DAX	Germany Stock Market Return
FTSE	UK Stock Market Return
NIK	Japan Stock Market Return
BVSP	Brazil Stock Market Return
EU	MSCI European Index Return
EM	MSCI Emerging Markets Index Return

Table 4. Error for the Istanbul stock exchange dataset. Gold standard trained with $N_{train} = 536$. Best results in bold.

	$N_{train} = 10$ (over 536)			$N_{train} = 30$ (over 536)		
	SBN	Credal	Belief Net	SBN	Credal	Belief Net
Actual RMSE	**0.131**	0.170	0.172	**0.088**	0.089	0.104
Predicted RMSE	0.146	0.223	0.124	0.093	0.140	0.068

5   Empirical Evaluation of User Co-design of Subjective Bayesian Networks

We selected three people among staff members and students at Cardiff University on the basis of their expertise with computational models of uncertainty. The first person should be considered an expert in probabilistic methods of inference; the second is a mature PhD student who spent many years in software companies and with a basic understanding of probabilistic methods of inference; the third is a sociologist with little or no experience with probabilistic methods of inference.

We prepared then a written briefing—see Appendix A—and an example of the mockup—analogous to Fig. 2. We then asked them to consider six cases, each of which refers to a specific command for an hypothetical conversational interface, namely:

1. explain, i.e. summarise the results;
2. explain ‹ company ›, i.e. describe the dependencies for a specific company, e.g. Henkel;
3. explain in detail ‹ company ›, i.e. describe *in detail*, i.e. with information on the probabilistic model—thus oriented to a more specialist audience—the dependencies for a specific company;
4. what happens if both ‹ company1 › and ‹ company2 › stock prices change?, i.e. exploring what-if scenarios;
5. what happens in detail if both ‹ company1 › and ‹ company2 › stock prices change?, i.e. exploring what-if scenarios for a more specialist audience;
6. ‹ company1 › depends on ‹ company2 ›, i.e. add a new dependency between two companies (cf. Fig. 2).

For each of those cases, we ask the participants to answer the following questions from the Subjective Usability Scale (SUS) questionnaire [4]. The numeric responses ranged between 1: Strongly Disagree, and 5: Strongly Agree. We asked the following questions:

Q1: I think that I would like to use this command frequently
Q2: I found the answer unnecessarily complex
Q3: I thought that the interaction was quite natural
Q4: I think that I would need the support of a technical person to be able to understand this interaction

At the end of the experiment, using the same numeric scale, participants were asked to answer additional questions from the SUS questionnaire:

SQ1: I think that I would like to use the conversational interface frequently
SQ2: I found the application unnecessarily complex
SQ3: I thought that overall the interaction was quite natural
SQ4: I think that overall I would need the support of a technical person to be able to understand this interaction
SQ5: I found the various commands well integrated
SQ6: I thought there was too much inconsistency in the application
SQ7: I would imagine that motivated users would learn to use the application very quickly
SQ8: I found the commands very cumbersome
SQ9: I think I would need to learn a lot of things before I could get going with the application

At any time, participants could write notes and comments in natural language to share with us. Moreover, the participants were reminded that their participation was entirely voluntary and that they were free to abandon the evaluation at any time with no need for explanations. This project received ethical approval from the School of Computer Science & Informatics Research Ethics Group.

5.1 Results

Explain. Figure 7 shows the mockup of interaction for the explain command and Fig. 12 depicts the results of the evaluation using error bars.

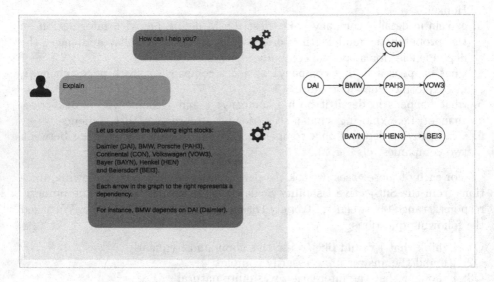

Fig. 7. Mockup for the explain command.

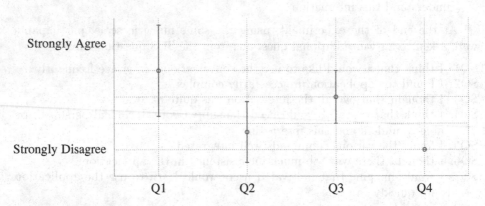

Fig. 8. Results of the evaluation for the explain command: average and standard deviation of the aggregated received answers. Collected data are not normally distributed. Q1: I think that I would like to use this command frequently. Q2: I found the answer unnecessarily complex. Q3: I thought that the interaction was quite natural. Q4: I think that I would need the support of a technical person to be able to understand this interaction

Explain ‹ Company ›. Figure 9 shows the mockup of interaction for the explain ‹ company › command and Fig. 10 depicts the results of the evaluation using error bars.

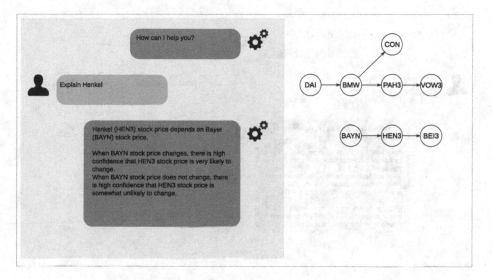

Fig. 9. Mockup for the explain ‹ company › command.

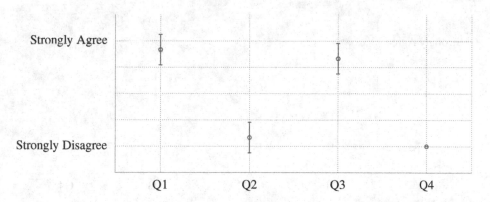

Fig. 10. Results of the evaluation for the explain ‹ company › command: average and standard deviation of the aggregated received answers. Collected data are not normally distributed. Q1: I think that I would like to use this command frequently. Q2: I found the answer unnecessarily complex. Q3: I thought that the interaction was quite natural. Q4: I think that I would need the support of a technical person to be able to understand this interaction

Explain in detail ‹ Company ›. Figure 11 shows the mockup of interaction for the explain in detail ‹ company › command and Fig. 12 depicts the results of the evaluation using error bars.

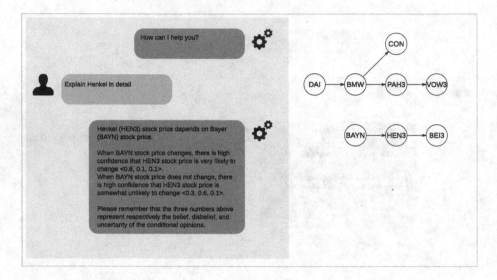

Fig. 11. Mockup for the explain in detail ‹ company › command.

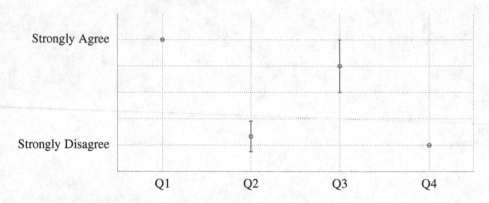

Fig. 12. Results of the evaluation for the explain in detail ‹ company › command: average and standard deviation of the aggregated received answers. Collected data are not normally distributed. Q1: I think that I would like to use this command frequently. Q2: I found the answer unnecessarily complex. Q3: I thought that the interaction was quite natural. Q4: I think that I would need the support of a technical person to be able to understand this interaction

What happens if ... Figure 13 shows the mockup of interaction for the what happens if ... command and Fig. 14 depicts the results of the evaluation using error bars.

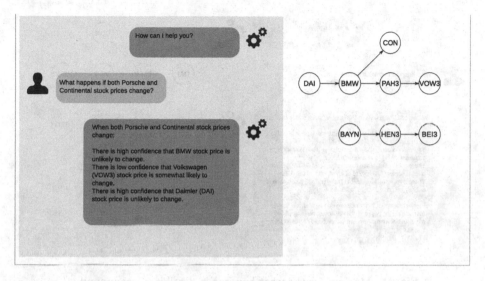

Fig. 13. Mockup for the what happens if ... command.

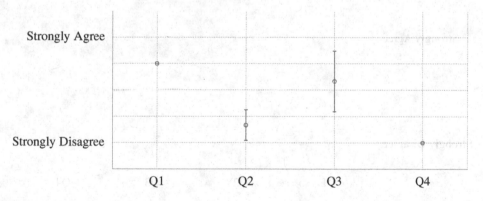

Fig. 14. Results of the evaluation for the what happens if ... command: average and standard deviation of the aggregated received answers. Collected data are not normally distributed. Q1: I think that I would like to use this command frequently. Q2: I found the answer unnecessarily complex. Q3: I thought that the interaction was quite natural. Q4: I think that I would need the support of a technical person to be able to understand this interaction

What happens in detail if ... Figure 15 shows the mockup of interaction for the what happens in detail if ... command and Fig. 16 depicts the results of the evaluation using error bars.

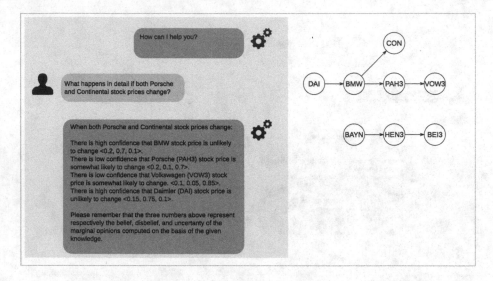

Fig. 15. Mockup for the what happens in detail if ... command.

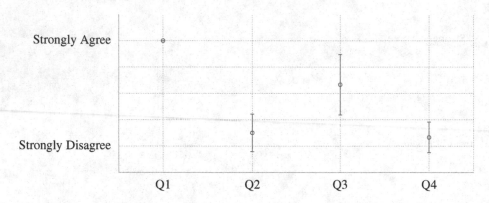

Fig. 16. Results of the evaluation for the what happens in detail if ... command: average and standard deviation of the aggregated received answers. Collected data are not normally distributed. Q1: I think that I would like to use this command frequently. Q2: I found the answer unnecessarily complex. Q3: I thought that the interaction was quite natural. Q4: I think that I would need the support of a technical person to be able to understand this interaction

⟨ **company1** ⟩ **depends on** ⟨ **company2** ⟩. Figure 2 shows the mockup of interaction for the ⟨ company1 ⟩ depends on ⟨ company2 ⟩ and Fig. 17 depicts the results of the evaluation using error bars.

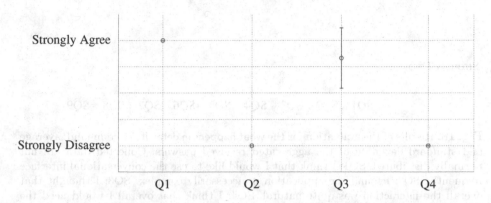

Fig. 17. Results of the evaluation for the ⟨ company1 ⟩ depends on ⟨ company2 ⟩ command: average and standard deviation of the aggregated received answers. Collected data are not normally distributed. Q1: I think that I would like to use this command frequently. Q2: I found the answer unnecessarily complex. Q3: I thought that the interaction was quite natural. Q4: I think that I would need the support of a technical person to be able to understand this interaction.

Overall Evaluation. Figure 18 depicts the results of the overall evaluation using error bars.

5.2   Summary of Evaluation

Considering the results summarised in Figs. 8, 9, 10, 11, 12, 13, 14, 15, 16, 17 and 18, we can conclude that in general our participants appreciated the idea of having a conversational interface for situational understanding. Indeed, the three participants mostly agreed with the positive statements and mostly disagreed with the negative statements in the questionnaire. There are, however, avenues for improvement.

As per the explain command, it is likely that users will use it only once, as their continuous interaction with the interface will lead them to familiarise with the graph-based interface. On this note, it would be useful to have multi-modal interaction, with also conditional probabilities tables associated with the graph with natural language labels or also subjective logic opinions. Also, looking at the collected data, it seems that providing the participants with additional information in textual format lead to a less "natural" interaction with the application. This might be correlated to the chosen scenario, or also to possible difficulties to understand SBNs: we will continue our investigation in the future taking these options into due consideration.

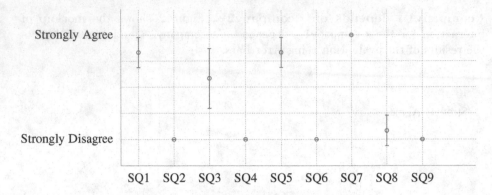

Fig. 18. Results of the evaluation for the what happens in detail if ... command: average and standard deviation of the aggregated received answers. Collected data are not normally distributed. SQ1: I think that I would like to use the conversational interface frequently. SQ2: I found the application unnecessarily complex. SQ3: I thought that overall the interaction was quite natural. SQ4: I think that overall I would need the support of a technical person to be able to understand this interaction. SQ5: I found the various commands well integrated. SQ6: I thought there was too much inconsistency in the application. SQ7: I would imagine that motivated users would learn to use the application very quickly. SQ8: I found the commands very cumbersome. SQ9: I think I would need to learn a lot of things before I could get going with the application.

Moreover, there is a missing command in the list, namely a help, which is probably the first command a fresh user will ask. Finally, it is unclear whether the supplied commands are sufficient for supporting all the tasks of an analyst: this specific point will require further analysis with realistic scenarios.

6   Conclusion

In this paper we presented a methodology to exploit human-machine coalitions for situational understanding, i.e., the ability to relate relevant information with dependencies and form logical conclusions as well as identifying gaps in information. This process requires the ability to reason inductively, for which one must exploit the machines' ability to learn from data, although important phenomena are often rare in occurrence, severely limiting the availability of instance data and hence the applicability of many machine learning approaches.

To this end, we discussed at length the benefits of SBNs, especially when training with sparse data, and in Sect. 4 we showed that they are superior to previous methods to reason over uncertain probabilistic networks, Credal networks and Belief Networks. In the future, we plan to compare SBN to other probabilistic models for dependencies such as Maximum Likelihood Estimation of an Alternating Renewal Process [18,21]. We considered two different datasets both related to the financial domain, but clearly SBNs can directly be applied to other datasets. We are working towards inference over general directed acyclic graphs as they characterise any joint probability distribution.

We also discussed the role that would be played by humans in situational understanding. Differently from other approaches aimed at explaining high-dimensional, multivariate feature spaces and dependencies to humans, e.g. [22], we believe a conversational interface like the one depicted in Fig. 2 can provide the right level of interactivity in the coalition of humans and machines for situational understanding. We are currently developing the first prototype of this conversational interface, and we are focusing on three major capabilities: (1) the ability to explain the dependencies (e.g., "When Bayer stock price changes, it is likely that. . ."); (2) the ability of what-if reasoning (e.g., "If Bayer stock price changes, then. . ."); and (3) as shown in Fig. 2, the ability to modify the dependency network. The preliminary evaluation we discussed in Sect. 5 suggests that conversational interfaces are a positive way to interact with complex decision making systems such as drawing inferences using Bayesian Networks. However, additional interfaces, including extending the graphical representation with conditional probabilities tables and enable their manipulation, need to be studied as they might suit some of the potential users. Moreover, evaluating the conversational interface in different case-studies might show which commands are mostly used, and eventually which ones need to implemented.

This opens a large spectrum of future work, including the ability to evaluate the human expertise and the quality of data. If a human user adds a dependency that is not supported by available data, it might suggest that the user has knowledge that is "out of scope" in the data and/or model. However, such an assertion may simply be erroneous or could indicate data quality issues such as data that are incomplete, biased, or corrupted.

A   Briefing Received by the Participants

A computer analysed the data of the German Stock Market Börse Frankfurt related to nine companies:

- Bayer, a pharmaceutical company
- Beiersdorf, a cosmetic company
- Henkel, a cosmetic company
- BMW, an automotive manufacturer
- Daimler, an automotive manufacturer
- Porsche, an automotive manufacturer
- Volkswagen, an automotive manufacturer
- Continental, a tyre manufacturer

In particular, the computer was programmed only to consider whether the closing value of a stock price was significantly different from same stock price at the closing time of the day before ($\pm 0.5\%$). And then the computer automatically derived possible dependencies between stocks.

Example. The Bayer stock value at the closing time on 7th December 2016 was 90.10; at the closing time on 8th December 2017 it was 93.17, thus with a significant change of 3.4%.

Similarly, the computer also analyses the changes of all the other companies considered in this study, thus producing a large table like the following:

Company	07/12/16	08/12/16	09/12/16	...
Bayer	Stable	Changed	Changed	...
Beiersdorf	Stable	Changed	Stable	...
Henkel	Stable	Stable	Stable	...
...	...	...	...	...

On the basis of such a large table, and by employing *Machine Learning* procedures, the computer identifies dependencies between companies' stock values. An example of such a dependencies can be:

When Bayer stock price changes, there is low confidence that Henkel stock price is unlikely to change.

References

1. Akbilgic, O., Bozdogan, H., Balaban, M.E.: A novel hybrid RBF neural networks model as a forecaster. Stat. Comput. **24**(3), 365–375 (2014)
2. Braines, D., Thomas, A., Kaplan, L., Sensoy, M., Ivanovska, M., Preece, A.D., Cerutti, F.: Human-in-the-loop situational understanding via subjective Bayesian networks. In: The 5th International Workshop on Graph Structures for Knowledge Representation and Reasoning (GKR 2017) (2017)
3. Brannon, N.G., Seiffertt, J.E., Draelos, T.J., Wunsch II, D.C.: Coordinated machine learning and decision support for situation awareness. Neural Netw. **22**(3), 316–325 (2009)
4. Brooke, J., et al.: SUS-A quick and dirty usability scale. In: Usability Evaluation in Industry, vol. 189(194), pp. 4–7 (1996)
5. Chawla, N.V., Japkowicz, N., Kotcz, A.: Editorial: special issue on learning from imbalanced data sets. ACM SIGKDD Explor. Newsl. **6**(1), 1–6 (2004)
6. Dhami, M.K., Mandel, D.R., Mellers, B.A., Tetlock, P.E.: Improving intelligence analysis with decision science. Perspect. Psychol. Sci. **10**(6), 753–757 (2015)
7. Dostal, B.C.: Enhancing situational understanding through employment of unmanned aerial vehicle. Army Transformation Taking Shape: Interim Brigade Combat Team Newsletter 01-18 (2007)
8. Helbing, D.: Globally networked risks and how to respond. Nature **497**(7447), 51–59 (2013)
9. Ivanovska, M., Jøsang, A., Kaplan, L., Sambo, F.: Subjective networks: perspectives and challenges. In: Croitoru, M., Marquis, P., Rudolph, S., Stapleton, G. (eds.) GKR 2015. LNCS (LNAI), vol. 9501, pp. 107–124. Springer, Cham (2015). https://doi.org/10.1007/978-3-319-28702-7_7
10. Jøsang, A.: Subjective Logic: A Formalism for Reasoning Under Uncertainty. Springer, Cham (2016). https://doi.org/10.1007/978-3-319-42337-1
11. Kahneman, D.: Thinking, Fast and Slow. Macmillan, Basingstoke (2011)

12. Kaplan, L., Ivanovska, M.: Efficient subjective Bayesian network belief propagation for trees. In: International Conference on Information Fusion (FUSION), pp. 1300–1307 (2016)
13. Kaplan, L., Ivanovska, M.: Efficient subjective Bayesian network belief propagation for singly-connected graphs. Int. J. Approx. Reason. (2017, submitted)
14. Kaplan, L., Şensoy, M., Chakraborty, S., de Mel, G.: Partial observable update for subjective logic and its application for trust estimation. Inf. Fusion **26**, 66–83 (2015)
15. Karlsson, A., Johansson, R., Andler, S.F.: An empirical comparison of Bayesian and credal networks for dependable high-level information fusion. In: International Conference on Information Fusion (FUSION), pp. 1–8 (2008)
16. LeCun, Y., Bengio, Y., Hinton, G.: Deep learning. Nature **521**(7553), 436–444 (2015)
17. Lerner, B., Malka, R.: Investigation of the K2 algorithm in learning Bayesian network classifiers. Appl. Artif. Intell. **25**(1), 74–96 (2011)
18. Lin, X., Moussawi, A., Korniss, G., Bakdash, J.Z., Szymanski, B.K.: Limits of risk predictability in a cascading alternating renewal process model. Sci. Rep. **7**(1), 6699 (2017)
19. Pearl, J.: Fusion, propagation, and structuring in belief networks. Artif. Intell. **29**(3), 241–288 (1986)
20. Smets, P.: Belief functions: the disjunctive rule of combination and the generalized Bayesian theorem. Int. J. Approx. Reason. **9**, 1–35 (1993)
21. Szymanski, B.K., Lin, X., Asztalos, A., Sreenivasan, S.: Failure dynamics of the global risk network. Sci. Rep. **5**, 10998 (2015)
22. Timmer, S.T., Meyer, J.J.C., Prakken, H., Renooij, S., Verheij, B.: A two-phase method for extracting explanatory arguments from Bayesian networks. Int. J. Approx. Reason **80**(C), 475–494 (2017)
23. Zaffalon, M., Fagiuoli, E.: 2U: an exact interval propagation algorithm for polytrees with binary variables. Artif. Intell. **106**(1), 77–107 (1998)
24. Zhou, Z.H., Sun, Y.Y., Li, Y.F.: Multi-instance learning by treating instances as Non-IID samples. In: Proceedings of the 26th Annual International Conference on Machine Learning, pp. 1249–1256. ACM (2009)

Counting and Conjunctive Queries in the Lifted Junction Tree Algorithm

Tanya Braun$^{(\boxtimes)}$ and Ralf Möller

Institute of Information Systems, Universität zu Lübeck, Lübeck, Germany
{braun,moeller}@ifis.uni-luebeck.de

Abstract. Standard approaches for inference in probabilistic formalisms with first-order constructs include lifted variable elimination (LVE) for single queries. To handle multiple queries efficiently, the lifted junction tree algorithm (LJT) uses a first-order cluster representation of a knowledge base and LVE in its computations. We extend LJT with a full formal specification of its algorithm steps incorporating (i) the lifting tool of counting and (ii) answering of conjunctive queries. Given multiple queries, e.g., in machine learning applications, our approach enables us to compute answers faster than the current LJT and existing approaches tailored for single queries.

1 Introduction

AI research and application areas such as natural language understanding and machine learning (ML) need efficient inference algorithms. Modeling realistic scenarios results in large probabilistic models that require reasoning about sets of individuals. Lifting uses symmetries in a model to speed up reasoning with known domain objects. We study the problem of reasoning in large models that exhibit symmetries. Our inputs are a model and queries for probabilities or probability distributions of random variables (randvars) given evidence. Inference tasks reduce to computing marginal distributions. We aim to enhance the efficiency of these computations when answering multiple queries, a common scenario in ML. We exploit that a model remains constant under multiple queries.

We have introduced a lifted junction tree algorithm (LJT) for multiple queries on models with first-order constructs [3]. LJT is based on the junction tree algorithm [17] and lifted variable elimination (LVE) as specified in [26]. LJT uses a first-order junction tree (FO jtree) to represent clusters of randvars in a model. This paper extends LJT and contributes the following: We give a formal specification of the LJT steps construction, message passing, and query answering. We incorporate counting as defined in [26] to lift more computations and allow a wider variety of model specifications. We adapt the LVE heuristic for elimination order for message passing and extend query answering for conjunctive queries that may cover multiple clusters based on [16].

LJT imposes some static overhead for building an FO jtree and message passing. Counting allows accelerating computations during message passing and

© Springer International Publishing AG, part of Springer Nature 2018
M. Croitoru et al. (Eds.): GKR 2017, LNAI 10775, pp. 54–72, 2018.
https://doi.org/10.1007/978-3-319-78102-0_3

query answering. Handling conjunctive queries allows for more complex queries. We significantly speed up runtime compared to LVE and LJT. Overall, we handle multiple queries more efficiently than approaches tailored for single queries.

The remainder of this paper has the following structure: First, we look at related work on exact lifted inference and the junction tree algorithm. Then, we introduce basic notations and data structures and recap LVE and LJT. We present our extension incorporating counting and conjunctive queries, followed by a brief empirical evaluation. Last, we present a conclusion and upcoming work.

2   Related Work

In the last two decades, researchers have sped up runtimes for inference significantly. Propositional formalisms benefit from variable elimination (VE) [28]. VE decomposes a model into subproblems to evaluate them in an efficient order. A decomposition tree (dtree) represents such a decomposition [9]. LVE, first introduced in [19] and expanded in [20], exploits symmetries at a global level. LVE saves computations by reusing intermediate results for isomorphic subproblems. Milch *et al.* introduce counting to lift certain computations where lifted summing out is not applicable [18]. Taghipour *et al.* extend the formalism to its current standard by generalising counting [26]. He formalises lifting by defining lifting operators. The operators appear in internal calculations of LJT.

For multiple queries in a propositional setting, Lauritzen and Spiegelhalter introduce junction trees (jtrees), a representation of clusters in a propositional model, along with a reasoning algorithm [17]. The algorithm distributes knowledge in a jtree with a message passing scheme, also known as probability propagation (PP), and answers queries on the smaller clusters. Shafer and Shenoy as well as Jensen *et al.* propose well known PP schemes [14,21]. They trade off runtime and storage differently, making them suitable for certain uses. Darwiche demonstrates a connection between jtrees and VE, namely, the clusters of a dtree form a jtree [10]. Taghipour *et al.* transfer the idea of dtrees to the first-order setting, introducing FO dtrees, allowing for a complexity analysis of lifted inference [25].

Lifted belief propagation (LBP) combines PP and lifting, often using lifted representations, e.g., with hyper-cubes [12,23]. Kersting and Ahmadi *et al.* present a counting LBP that runs a colouring algorithm with additional mechanisms for dynamic models [1,15]. To the best of our knowledge, none of them use jtrees to focus on multiple queries.

Lifted inference sparks progress in various fields. Van den Broeck applies lifting to weighted model counting [5] and first-order knowledge compilation, with newer work on asymmetrical models [6]. To scale lifting, Das *et al.* use graph data bases storing compiled models to count faster [11]. Both works are interesting avenues for future work. Chavira and Darwiche focus on knowledge compilation as well also addressing the setting of multiple queries and using local symmetries [7]. Other areas incorporate lifting to enhance efficiency, including continuous or dynamic models [8,27], logic programming [2], and theorem proving [13].

We apply lifting to jtrees, introducing FO jtrees and provide LJT as a reasoning algorithm using LVE as a subroutine [3]. Currently, LJT does not include counting and handling of conjunctive queries. We widen the scope of the algorithm with our extension and speed up inference time.

3   Preliminaries

This section introduces basic notations, the FO dtree and FO jtree data structures, and recaps LVE and LJT based on [3,26]. We assume familiarity with common notions such as jtrees and dtrees (for an introduction, see, e.g., [10]).

3.1   Parameterised Models

Parameterised models compactly represent models with first-order constructs using logical variables (logvars) as parameters. We begin with denoting basic blocks on our way to build a full model.

Definition 1. *Let* **L** *be a set of logvar names,* Φ *a set of factor names, and* **R** *a set of randvar names. A parameterised randvar* (PRV) $R(L_1, \ldots, L_n), n \geq 0$, *is a syntactical construct of a randvar* $R \in$ **R** *combined with logvars* $L_1, \ldots, L_n \in$ **L** *to represent a set of randvars that behave identically. Each logvar* L *has a domain, denoted by* $\mathcal{D}(L)$. *The term* $range(A)$ *denotes the possible values of some PRV* A. *A constraint* $(\mathbf{X}, C_{\mathbf{X}})$ *is a tuple with a sequence of logvars* $\mathbf{X} = (X_1, \ldots, X_n)$ *and a set* $C_{\mathbf{X}} \subseteq \times_{i=1}^{n} \mathcal{D}(X_i)$. C *allows for restricting logvars to certain domain values. The symbol* $\top$ *marks that no restrictions apply and may be omitted.*

The terms $lv(P)$ and $rv(P)$ refer to the logvars and PRVs with constraints, respectively, in some P. The term $gr(P)$ denotes the set of instances of P with all logvars in P grounded w.r.t. constraints or domains. Let us look at an example.

Example 1. We model that people attend conferences and do research on some topic depending on whether this topic is considered hot. Later, we want to encode that, e.g., the potential increases that a person does research on some topic if this topic is hot. The potential should be identical for different people. So, we use PRVs to represent a set of people, e.g., *alice*, *eve*, and *bob*, that have the same behaviour for some randvar.

Given randvar names $HoTpc$, $AttCnf$, and Res and logvar name X, we build PRVs $HoTpc$, $AttCnf(X)$, and $Res(X)$. The domain of X is given by $\mathcal{D}(X) = \{alice, eve, bob\}$. Each PRV has the range $\{true, false\}$. $HoTpc$ is not parameterised and represents a propositional randvar, while $AttCnf(X)$ and $Res(X)$ represent sets of randvars. A constraint can modify which randvars $AttCnf(X)$ and $Res(X)$ represent. A constraint $C = \top$ for X does not restrict X. A constraint $C' = (X, \{alice, eve\})$ restricts X to not take the value *bob*. $gr(Res(X)|C')$ contains $Res(alice)$ and $Res(eve)$ while $gr(Res(X)|C)$ also contains $Res(bob)$.

We use the basic blocks of logvars, PRVs, and constraints to form more complex structures that make up a model.

Definition 2. *A parametric factor* (parfactor) *g consists of a function mapping argument values to real values. We denote a parfactor by $\forall \mathbf{X} : \phi(\mathcal{A}) \mid C$ where $\mathbf{X} \in \mathbf{L}$ is a set of logvars that the factor generalises over. $\mathcal{A} = (A_1, \ldots, A_n)$ is a sequence of PRVs. Each PRV is built from $\mathbf{R}$ and possibly $\mathbf{X}$. We omit $(\forall \mathbf{X} :)$ if $\mathbf{X} = lv(\mathcal{A})$. $\phi : \times_{i=1}^{n} range(A_i) \mapsto \mathbb{R}^{+}$ is a function with name $\phi \in \Phi$, identical for all instances of $\mathcal{A}$. A full specification of ϕ requires listing all input-output values. C is a constraint on $\mathbf{L}$. A set of parfactors forms a model $G := \{g_i\}_{i=1}^{n}$. G represents the probability distribution $P_G = \frac{1}{Z} \prod_{f \in gr(G)} \phi_f(\mathcal{A}_f)$ with Z as the normalisation constant.*

For our above example of *alice*, *eve*, and *bob* doing research and attending conferences influenced by a topic considered hot, we can build a parfactor that encodes this identical behaviour.

Example 2. With the above PRVs and a factor name ϕ, we build a parfactor $g = \phi(HoTpc, AttCnf(X), Res(X)) \mid \top$. The mappings with randomly chosen potentials are (with *true* = 1 and *false* = 0):

$$(0,0,0) \rightarrow 10, \quad (0,0,1) \rightarrow 3, \quad (0,1,0) \rightarrow 3, \quad (0,1,1) \rightarrow 7,$$
$$(1,0,0) \rightarrow 6, \quad (1,0,1) \rightarrow 6, \quad (1,1,0) \rightarrow 5, \quad (1,1,1) \rightarrow 9$$

The $\top$ constraint means ϕ holds for *alice*, *eve*, and *bob*. $gr(g)$ contains three factors with identical potential functions.

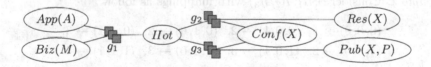

Fig. 1. Parfactor graph for G_{ex}

We compile a model G_{ex} building on Example 2: The topic allows for business markets and application areas and for a person to publish in publications. Logvars encode that there are several markets (M), areas (A), and publications (P).

Example 3. Let $\mathbf{L} = \{A, M, P, X\}$, $\Phi = \{\phi_1, \phi_2, \phi_3\}$, and $\mathbf{R} = \{HoTpc, Biz, App, AttCnf, Res, Pub\}$. The domains for the logvars are $\mathcal{D}(A) = \{ml, nlp\}$, $\mathcal{D}(M) = \{itsec, ehealth\}$, $\mathcal{D}(P) = \{p_1, p_2\}$, and $\mathcal{D}(X) = \{alice, eve, bob\}$. In addition to $HoTpc$, $AttCnf(X)$, and $Res(X)$, we build the binary PRVs $Biz(M)$, $App(A)$, and $Pub(X, P)$. The model reads $G_{ex} = \{g_1, g_2, g_3\}$,

- $g_1 = \phi_1(HoTpc, App(A), Biz(M)) \mid C_1$,
- $g_2 = \phi_2(HoTpc, AttCnf(X), Res(X)) \mid C_2$, and
- $g_3 = \phi_3(HoTpc, AttCnf(X), Pub(X, P)) \mid C_3$.

We omit concrete functions for ϕ_1, ϕ_2, and ϕ_3 at this point. C_1, C_2, and C_3 are $\top$ constraints, meaning ϕ_1, ϕ_2, and ϕ_3 apply for all possible tuples. Figure 1 depicts G_{ex} as a graph with six variable nodes for the PRVs and three factor nodes for g_1, g_2, and g_3 with edges to the PRVs involved.

The semantics of a model is given by grounding and building a full joint distribution. The query answering (QA) problem asks for a probability distribution of a randvar w.r.t. a model's joint distribution and fixed events (evidence). Formally, $P(Q|\mathbf{E})$ denotes a query where Q is a grounded PRV (a normal randvar) and $\mathbf{E}$ is a set of events (grounded PRVs with fixed range values). A query for G_{ex} is $P(Pub(eve, p_1)|AttCnf(eve) = true)$, with $AttCnf(eve) = true$ a fixed event of eve attending conferences and asking for the probability distribution of eve publishing in p_1. Next, we look at algorithms for QA. They seek to avoid grounding as well as building a full joint distribution.

3.2 Lifted Variable Elimination

LVE employs two main techniques for QA, namely (i) decomposition into isomorphic subproblems and (ii) counting of domain values leading to a certain range value of PRV given the remaining PRVs in a parfactor. The first technique refers to lifted summing out. The idea is to compute VE for one case and then exponentiate the result with the number of isomorphic instances.

The second technique, counting, exploits that all instances of a PRV A evaluate to $range(A)$. A counting randvar (CRV) encodes for n interchangeable randvars, i.e., instances of A, how many have a certain value.

Example 4. Consider $\phi(R_1, R_2, R_3)$ with mappings as follow:

$$(0,0,0) \rightarrow 1, \quad (0,0,1) \rightarrow 2, \quad (0,1,0) \rightarrow 2, \quad (0,1,1) \rightarrow 3,$$
$$(1,0,0) \rightarrow 2, \quad (1,0,1) \rightarrow 3, \quad (1,1,0) \rightarrow 3, \quad (1,1,1) \rightarrow 4$$

The potentials for $(0,0,1)$, $(0,1,0)$, and $(1,0,0)$ are identical, namely 2, and the argument values exhibit that two of them are *false* and one is *true*. The same observation holds for two arguments being *true* and one *false*, all mapping to the potential of 3. So, instead of using eight mappings, we use a histogram to encode how many of the R randvars have a specific value that maps to the corresponding potential (first position $R = 1$, second $R = 0$):

$$[0,3] \rightarrow 1, \quad [1,2] \rightarrow 2, \quad [2,1] \rightarrow 3, \quad [3,0] \rightarrow 4$$

To refer to a set of randvars in this counted version, we use a CRV.

Definition 3. *We denote a CRV by $\#_{X \in C}[P(\mathbf{X})]$ for a PRV $P(\mathbf{X})$ and constraint C, where $lv(\mathbf{X}) = \{X\}$ (meaning all other inputs are constant). The range of a CRV is the space of possible histograms. Since counting binds logvar X, $lv(\#_{X \in C}[P(\mathbf{X})]) = \mathbf{X} \setminus \{X\}$. A histogram h is a set of tuples $\{(v_i, n_i)\}_{i=1}^{m}$,*

$m = |range(P(\mathbf{X}))|$, $n_i \in \mathbb{N}$, and $\sum_i n_i = |gr(P(\mathbf{X})|C)|$. *A shorthand notation is* $[n_1, \ldots, n_m]$. $h(v_i)$ *returns* n_i. *If* $\{X\} \subset lv(\mathbf{X})$, *the CRV is a parameterised CRV (PCRV) and represents a set of CRVs. We count-convert a logvar* X *in a PRV* $A_i \in \mathcal{A}$ *in a parfactor* $\mathbf{L} : \phi(\mathcal{A})|C$ *leading to a CRV* A_i'. *In the new parfactor,* ϕ' *has a histogram* h *as input for* A_i'. $\phi'(\ldots, a_{i-1}, h, a_{i+1}, \ldots)$ *maps to* $\prod_{a_i \in range(A_i)} \phi(\ldots, a_{i-1}, a_i, a_{i+1}, \ldots)^{h(a_i)}$.

The techniques have preconditions [26], e.g., to sum out PRV A in parfactor g, $lv(A) = lv(g)$. To count-convert logvar X in g, only one input in g contains X. Counting binds X, i.e., $lv(\#_{X \in C}[P(\mathbf{X})]) = \mathbf{X} \setminus \{X\}$, possibly allowing summing out another PRV that we otherwise need to ground. LVE includes further techniques to enable lifted summing out. Grounding is its last resort where it replaces a logvar with each value in a constraint, duplicating the affected parfactors. To eliminate a next PRV, LVE chooses from operations applicable to the model based on the size of the intermediate result after applying an operation. Let us apply LVE to $g_1 \in G_{ex}$.

Example 5. In $\phi_1(HoTpc, App(A), Biz(M))$, we cannot sum out any PRV as neither includes both logvars. One may ground M, leading to $|gr(M)|$ parfactors of the form $\phi(HoTpc, App(A), Biz(m))$ for all $m \subset gr(M)$. To eliminate $App(A)$, one multiplies all new parfactors into one with $HoTpc$, $App(A)$, and all instances of $Biz(M)$ as arguments. All randvars represented by $Biz(M)$ lead to true or false and we can count them as in Example 4. We count convert to avoid the grounding step. We can rewrite $Biz(M)$ into $\#_M[Biz(M)]$ and g_1 into $g_1' = \phi'(HoTpc, App(A), \#_M[Biz(M)])|C_1$. The CRV refers to histograms that specify for each value $v \in range(Biz(M))$ how many grounded PRVs evaluate to v. Given the previous mappings $(hot, app, true) \mapsto x$ and $(hot, app, false) \mapsto y$ in ϕ, ϕ' maps $(hot, app, [n_1, n_2])$ to $x^{n_1} y^{n_2}$. Since M is no longer a regular logvar, we sum out $App(A)$ using standard VE and exponentiate the result with $|gr(A)| = 2$.

3.3 FO Dtrees

VE recursively decomposes a model into partitions that include randvars not part of any other partition. A dtree represents these decompositions. With lifting, a dtree needs to represent isomorphic instances as well. We do so by grounding a subset of the model logvars with representative objects, called decomposition into partial groundings (DPG; requires a normal form, see [26]). In an FO dtree, DPG nodes represent such DPGs.

Definition 4. *A DPG node* $T_{\mathbf{X}}$ *is given by a 3-tuple* $(\mathbf{X}, \mathbf{x}, C)$ *where* $\mathbf{X} = \{X_1, \ldots X_k\}$ *is a set of logvars of the same domain* $\mathcal{D}_{\mathbf{X}}$, $\mathbf{x} = \{x_1, \ldots x_k\}$ *is a set of representative objects from* $\mathcal{D}_{\mathbf{X}}$, *and* C *is a constraint on* $\mathbf{x}$ *such that* $\forall i, j : x_i \neq x_j$. *We label* $T_{\mathbf{X}}$ *by* $(\forall \mathbf{x} : C)$ *in the FO dtree.* $T_{\mathbf{X}}$ *has a child* $T_{\mathbf{x}}$. *The decomposed model at* $T_{\mathbf{x}}$ *is a representative of* $T_{\mathbf{X}}$ *using a substitution* $\theta = \{X_i \rightarrow x_i\}_{i=1}^k$ *mapping* $\mathbf{X}$ *to* $\mathbf{x}$.

With a means to represent isomorphic instances and as such, lifted summing out in a dtree, we define FO dtrees for decompositions of a model during LVE.

Definition 5. *An FO dtree for a model G is a tree in which (i) non-leaf nodes can be DPG nodes, (ii) each leaf contains a factor (parfactor with representative objects), (iii) each leaf with representative object x descends from exactly one DPG node T_X such that $x \in \mathbf{x}$, (iv) each leaf descending from DPG node T_X has all representative objects $\mathbf{x}$ in its factor, and (v) for each DPG node T_X, $\mathbf{X} = \{X_1, \ldots X_k\}$, T_X has $k!$ children $\{T_i\}_{i=1}^{k!}$, which are isomorphic up to permutation of $\mathbf{x}$. All leaf factors combined correspond to G.*

The clusters of an FO dtree form an FO jtree. We compute clusters analogously to ground dtrees. A *cluster* of a node T is the union of its cutset and context. A cutset is the set of randvars shared between any two children minus the randvars in any ancestor cutset. A context is the intersection of its randvars and those in any ancestor cutset. We can count-convert logvars $\mathbf{X}$ if, at DPG node T_X, $\mathbf{X}$ appear in the cluster at T_X. Next, we inspect an FO dtree for G_{ex}.

Example 6. Figure 2 depicts an FO dtree without set braces and $\top$ constraints. The root partitions G_{ex} based on logvars with children $T_A = (A, a, \top)$ and $T_X = (X, x, \top)$. The models of both children share randvar $HoTpc$ while the other PRVs appear in only one of them. T_A has a child T_a with model $\{g_1' = \phi_1'(HoTpc, App(a), Biz(M))\}$, representative object a replacing A. Child node $T_M = (M, m, \top)$ has a child T_m with model $\{g_1'' = \phi_1''(HoTpc, App(a), Biz(m))\}$. g_1'' is ground so we have a leaf node. T_X has a child T_x with the model $\{g_2' = \phi_2'(HoTpc, AttCnf(x), Res(x)), g_3 = \phi_3(HoTpc, AttCnf(x), Pub(x, P))\}$. The children are a leaf node for g_2' and a node $T_P = (P, p, \top)$ with child T_p and model $\{g_3'' = \phi_3''(HoTpc, AttCnf(x), Pub(x, p))\}$. g_2' includes randvar $Res(x)$ not part of the model under T_P, which in return contains $Pub(x, P)$. T_p has a leaf child for g_3''. The PRVs pinned to inner nodes are clusters. Leaf clusters consist of factor arguments. As M appears in the cluster of T_M, it is count-convertible.

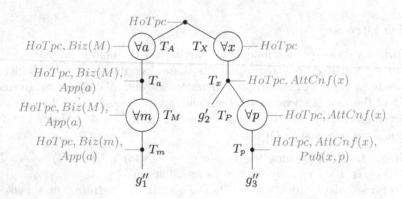

Fig. 2. FO dtree for G_{ex} (clusters for inner nodes in gray)

3.4   FO Jtrees

LJT runs on FO jtrees using logvars to encode symmetries in FO dtree clusters. We define a parameterised cluster (parcluster), i.e., a set of PRVs.

Definition 6. *A parcluster* **C** *is denoted by* $\forall$**L** : **A** | C *where* **L** *is a set of logvars and* **A** *is a set of PRVs with* $lv(\mathbf{A}) \subseteq \mathbf{L}$. *We omit* ($\forall$**L** :) *if* $\mathbf{L} = lv(\mathbf{A})$. *Constraint* C *puts limitations on logvars and representative objects. LJT assigns the parfactors of the input model to parclusters. A parfactor* $\phi(\mathbf{A}_\phi)|C_\phi$ *assigned to* $\mathbf{C}_i$ *at node* i *must fulfil (i)* $\mathbf{A}_\phi \subseteq \mathbf{A}$, *(ii)* $lv(\mathbf{A}_\phi) \subseteq \mathbf{L}$, *and (iii)* $C_\phi \subseteq C$. *We call the set of assigned parfactors a local model* G_i.

Next, we define FO jtrees, analogous to propositional jtrees, with parclusters replacing clusters and parfactors replacing factors.

Definition 7. *An FO jtree for a model* G *is a pair* $(\mathcal{J}, f_\mathbf{C})$ *where* $\mathcal{J}$ *is a cycle-free graph and* $f_\mathbf{C}$ *is a function mapping each node* i *in* $\mathcal{J}$ *to a label* $\mathbf{C}_i$ *called a parcluster. An FO jtree must satisfy three properties: (i) A parcluster* $\mathbf{C}_i$ *is a set of PRVs from* G. *(ii) For every parfactor* $g = \phi(\mathcal{A})|C$ *in* G, $\mathcal{A}$ *appears in some* $\mathbf{C}_i$. *(iii) If a PRV from* G *appears in* $\mathbf{C}_i$ *and* $\mathbf{C}_j$, *it must appear in every parcluster on the path between nodes* i *and* j *in* $\mathcal{J}$. *Parameterised set* $\mathbf{S}_{ij}$, *called separator of edge* i—j *in* $\mathcal{J}$, *contains the shared randvars of* $\mathbf{C}_i$ *and* $\mathbf{C}_j$.

An FO jtree is *minimal* if it ceases to be one if removing a PRV from any parcluster. The clusters of an FO dtree form a non-minimal FO jtree. To minimise, we merge neighbouring nodes if one parcluster is a subset of the other.

3.5   Lifted Junction Tree Algorithm

LJT provides an efficient way for answering a set of queries $\{Q_i\}_{i=1}^m$ given a model G. The main workflow is: (i) Construct an FO jtree for G. (ii) Pass messages. (iii) Compute answers for $\{Q_i\}_{i=1}^m$. For details regarding evidence, see [4].

FO jtree construction uses the clusters of an FO dtree for G. Message passing distributes local information at nodes to the other nodes. Two passes propagating information from the periphery to the inner nodes and back suffice [17]. LJT uses LVE to calculate the content of a message based on separators. If a node has received messages from all neighbours but one, it sends a message to the remaining neighbour (*inbound* pass). In the *outbound* pass, messages flow in the opposite direction. A query asks for the probability distribution (or the probability of a value) of a single grounded PRV, the query term. For each query, LJT finds a node whose parcluster contains the query term and sums out all non-query terms in its parfactors and received messages.

Since we extend LJT, we provide more details on the individual steps and an example in the next section.

4   Extended Lifted Junction Tree Algorithm

We extend LJT formally specifying its steps, incorporating counting and conjunctive queries. Algorithm 1 provides an outline of LJT.

Algorithm 1. Lifted Junction Tree Algorithm

---

function FOJT(Model G, Queries $\{\mathbf{Q}_i\}_{i=1}^m$)
    FO jtree $\mathcal{J}$ = FO-JTREE(G)
    PASSMESSAGES($\mathcal{J}$)
    GETANSWERS($\mathcal{J}$,$\{\mathbf{Q}_i\}_{i=1}^m$)

4.1   Construction

LJT constructs an FO jtree using FO dtree clusters. The FO dtree is constructed using a naive algorithm proposed in [24] that splits a model based on logvars if no DPG is possible. [24] also provides how to calculate clusters. We formalise how we convert the clusters into parclusters and when and how merging proceeds.

A cluster $\mathbf{A}_i$ of an FO dtree node i forms a parcluster $\forall \mathbf{L} : \mathbf{A} | C$ with

- $\mathbf{A} = \mathbf{A}_i$,
- $\mathbf{L} = lv(\mathbf{A}_i)$, if i is a DPG node $(\mathbf{X}, \mathbf{x}, C_{\mathbf{X}})$, then $\mathbf{L} = lv(\mathbf{A}_T) \cup \mathbf{X}$,
- $C = \emptyset$, if i is a DPG node $(\mathbf{X}, \mathbf{x}, C_{\mathbf{X}})$, then $C = C_{\mathbf{X}}$, and
- $G_i = \emptyset$, if i is a leaf node with factor g, then $G_i = \{g\}$.

For *merging*, we need set relations and operations. A parcluster $\mathbf{C}_i$ is a *subset* of parcluster $\mathbf{C}_j$, denoted by $\mathbf{C}_i \subseteq \mathbf{C}_j$, iff $gr(\mathbf{C}_i) \subseteq gr(\mathbf{C}_j)$. Exploiting that parclusters have certain properties by way of construction (e.g., domains are either distinct or identical), we need not ground but check parclusters component-wise. Other relations and operations are defined analogously.

Parclusters $\mathbf{C}_i$ and $\mathbf{C}_j$ with local models G_i and G_j are mergeable if $\mathbf{C}_i \subseteq \mathbf{C}_j \vee \mathbf{C}_j \subseteq \mathbf{C}_i$. The merged parcluster $\mathbf{C}_k$ and its local model G_k are given by $\mathbf{C}_k = \mathbf{C}_i \cup \mathbf{C}_j$ and $G_k = G_i \cup G_j$. The new node k takes over all neighbours of i and j. If we merge two parcluster, one with logvars $\mathbf{X}$ and one with representative objects $\mathbf{x}$, we first perform the inverse of substitution $\theta = \{X_i \rightarrow x_i\}_{i=1}^k$, performed at DPG node $T_{\mathbf{X}}$ in the underlying FO dtree, mapping $\mathbf{x}$ back onto $\mathbf{X}$.

Example 7. After converting the clusters in Fig. 2 into parclusters, we look at the leaf node with local model $\{g_3''\}$. As the neighbouring parcluster is identical, we merge them. Keeping them separate would mean sending a message with g_3'' leading to two nodes with identical information. We merge the next neighbour as well but replacing p with P again. Merging continues until we reach the node corresponding to the root in the FO dtree. The node with g_2' in its local model does not merge since its parcluster includes PRV $Res(x)$ (but applies $x \mapsto X$). The same procedure iteratively merges the nodes containing PRVs $HoTpc$, $App(A)$, and $Biz(M)$. Figure 3 shows the final result with three parclusters,

- $\mathbf{C}_1 = \forall A, M : \{HoTpc, App(A), Biz(M)\} | \top$,
- $\mathbf{C}_2 = \forall X : \{HoTpc, AttCnf(X), Res(X)\} | \top$, and
- $\mathbf{C}_3 = \forall X, P : \{HoTpc, AttCnf(X), Pub(X, P)\} | \top$.

$\mathbf{S}_{12} = \mathbf{S}_{21} = \{HoTpc\}$ and $\mathbf{S}_{23} = \mathbf{S}_{32} = \{HoTpc, AttCnf(X)\}$ are the separators. Each local model consists of one parfactor, which is not a common scenario.

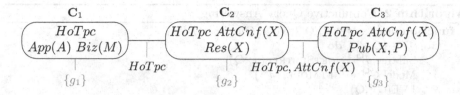

Fig. 3. FO jtree for G_{ex} (parcluster models in gray)

Regarding the extensions, *conjunctive queries* do not have any influence on construction. *Counting* affects construction w.r.t. PCRVs. We allow PCRVs in the input model, increasing its expressivity. PCRVs facilitate specifying counting behaviour explicitly in the model description. Construction handles PCRVs along with PRVs, becoming part of parclusters and possibly separators. Though we can identify logvars for count conversion in the FO dtree, we do not use this feature as explained in the next subsection.

Given an FO jtree for an input model, the next step in LJT is message passing, which we discuss next.

4.2   Message Passing

Message passing starts at the periphery, moves inwards, and then in the opposite direction to distribute all local information through the whole FO jtree. We define a message, discuss the effects of the extensions on messages, and as a consequence of counting, adapt the LVE heuristic selecting the next operation for calculating a message.

For a message from node i to node j, LJT encodes information present at i in parfactors over separator S_{ij} since j can process the PRVs in S_{ij}. Formally, a *message* m_{ij} from i with parcluster C_i and local model G_i to j is a set of parfactors, each with a subset of S_{ij} as arguments. To calculate m_{ij}, LJT eliminates all PRVs not in S_{ij} from G_i and the messages from all other neighbours using LVE, as described by

$$m_{ij} = \sum_{E \in \mathbf{E}_i} \prod_{g \in G'} g, \ \mathbf{E}_i = \mathbf{C}_i \setminus \mathbf{S}_{ij}, \ G' = G_i \cup \{m_{ik}\}_{k \neq j}.$$

m_{ij} can be a set of parfactors as LVE only multiplies parfactors if necessary. Let us look at messages in the FO jtree for G_{ex}.

Example 8. In the FO jtree for G_{ex} as depicted in Fig. 3, messages flow from nodes 1 and 3 to node 2 and back. Messages between nodes 1 and 2 have the argument $HoTpc$, messages between nodes 2 and 3 the arguments $HoTpc$ and $AttCnf(X)$. Inbound, the messages are m_{12} and m_{32}. For m_{12}, LJT eliminates $\mathbf{E}_1 = \{App(A), Biz(M)\}$ from $G' = G_1$ as in Example 5. For m_{32}, LJT eliminates $\mathbf{E}_3 = \{Pub(X, P)\}$ from $G' = G_3$ using lifted summing out on $Pub(X, P)$. At this point, node 2 has all information in the model in its local model and received

---

Algorithm 2. Conjunctive Query Answering

function GETANSWERS(FO Jtree J, Queries $\{\mathbf{Q}_i\}_{i=1}^m$)
    for $\mathbf{Q} \in \{Q_i\}_{i=1}^m$ **do**
        Subtree $J^{\mathbf{Q}} \leftarrow$ GETSUBTREE(J, $\mathbf{Q}$)
        Model $G^{\mathbf{Q}} \leftarrow$ GETMODEL($J^{\mathbf{Q}}$)
        LVE($G^{\mathbf{Q}}$,$\mathbf{Q}$)

---

messages, encoded in its parcluster PRVs. Outbound, node 2 propagates this information to node 1 with message m_{21} and to node 3 with message m_{23}. For m_{21}, LJT sums out $\mathbf{E}_2 = \{AttCnf(X), Res(X)\}$ from $G' = F_2 \cup \{m_{32}\}$ and for m_{23}, $\mathbf{E}_2 = \{Res(X)\}$ from $G' = F_2 \cup \{m_{12}\}$.

The extensions again only influence LJT on behalf of counting since *conjunctive queries* do not affect message passing which is independent of any queries. As mentioned above, *counting* appears in the form of PCRVs in an input model. As part of a model, PCRVs appear during message passing in a separator or in the set of PRVs to eliminate and are handled accordingly.

We do not count-convert logvars identified for count conversion in the FO dtree. Consider a scenario where PRVs $App(A)$ and $Biz(M)$ are in a parcluster and $App(A)$ in one separator. Assume given an FO dtree, we converted $Biz(M)$ into a CRV. Then, we still need to count-convert $App(A)$ to sum out $Biz(M)$, making the count conversion of logvar M superfluous. Since we cannot always determine from the clusters in the FO dtree if count conversion is reasonable for message passing, we do not count-convert in the FO dtree.

Example 8 references another use of counting, namely, as a means to enable a sum-out operation after count conversion when calculating a message. Without counting, the algorithm would need to ground a logvar. After count conversion, the new PCRV becomes part of the model that is used for further calculating the message. Then, the scenario plays out as described above when PCRVs are part of the model itself. The new PCRV either needs to be eliminated or becomes part of the message if the original PRV is part of the separator. If the new PCRV is part of the message, it becomes part of message calculations at the receiver.

The *heuristic* LVE uses no longer works for LJT in all cases. Consider the scenario from before with PRVs $App(A)$ and $Biz(M)$ in a parcluster and $App(A)$ in a separator. Using counting conversion on A, we can sum out $Biz(M)$. Assume that A has 50 domain values while M has 10. LVE would count-convert M as it leads to a smaller parfactor than count-converting A. After the count conversion, it still cannot sum out $\#_M[Biz(M)]$. So, it count-converts A to finally sum out $\#_M[Biz(M)]$, making the first count conversion unnecessary. For LJT, we require the heuristic to consider the PRVs in a separator.

We adapt the heuristic by dividing applicable counting operations into one part with operations for PRVs to eliminate and another part with operations for separator PRVs. If the operation with the lowest cost comes from the first part, we select the cheapest operation from the second part if not empty. With the adapted heuristic, we save superfluous applications of LVE operators.

After receiving messages from each neighbour, the parclusters hold in their local models all information to answer queries on its PRVs.

4.3 Query Answering

While *conjunctive queries* so far do not change LJT, query answering changes as we allow for multiple grounded PRVs $\mathbf{Q}$ in a query. Since we do not discuss evidence, a query has the form $P(\mathbf{Q})$ with a set of grounded PRVs $\mathbf{Q}$. LJT has as input a set of queries that now can each be a set of grounded PRVs $\mathbf{Q}_i$ instead of a single grounded PRV Q_i.

Query answering so far has meant to find a parcluster that contains the query randvar Q_i and eliminate all non-query PRVs from its local model using LVE. With a set of grounded PRVs in a query, we may have query randvars that are not part of one parcluster. We could force LJT to build an FO jtree with all query randvars in one parcluster but the forced construction inhibits fast query answering for other queries. Additionally, it assumes that we know a query in advance. Hence, we adapt the idea of so called out-of-clique inference [16], which extracts necessary information per query from a standard jtree.

Algorithm 2 shows a pseudo code description of our approach. We find a subtree of the FO jtree that covers all query randvars $\mathbf{Q}$. From the parclusters in the subtree, we extract a model to answer $\mathbf{Q}$ with LVE handling multiple query randvars. A more detailed description of each step follows, starting with identifying a subtree.

Subtree Identification. The goal is to find a subtree of the FO jtree where the subtree parclusters cover all query randvars $\mathbf{Q}$. Since the subtree is the basis for model extraction, the subtree should result in the smallest model possible in terms of number of PRVs. In a straightforward way, LJT finds a first node that covers at least part of $\mathbf{Q}$ and uses it as the first node in the subtree $J^{\mathbf{Q}}$. Then, it adds further nodes that cover still missing query randvars closest to the current $J^{\mathbf{Q}}$. Future work includes ways of finding a reasonably small subtree efficiently.

From the subtree covering all query randvars, LJT needs to extract a model to actually answer the query.

Model Extraction. We build a model $G^{\mathbf{Q}}$ from subtree $J^{\mathbf{Q}}$. The extracted model may not contain any duplicate information, meaning we cannot simply use all local models and messages within $J^{\mathbf{Q}}$. Instead, we use the local models at the nodes in $J^{\mathbf{Q}}$ and the messages that the nodes at the borders of $J^{\mathbf{Q}}$ received from outside $J^{\mathbf{Q}}$. Since LJT assigns each parfactor in G to exactly one parcluster, the local models hold no duplicate information. The border messages store all information from outside the subtree. Ignoring the messages within the subtree, we do not duplicate information through a message.

The remaining step in answering a query is to let LVE answer the query using the extracted model.

Query Answering. Using the model $G^\mathbf{Q}$ built during model extraction, LJT performs LVE to answer a query over the randvars $\mathbf{Q}^\mathbf{Q}$. Though LVE as described by [26] does not explicitly mention conjunctive queries, the formalism allows for multiple query randvars.

Query answering needs an operation called shattering that splits the parfactors in a model based on query randvars. For one query randvar Q, a split means we add a duplicate of each parfactor that covers Q and use the constraint to restrict the PRV in one parfactor to Q and the other to the remaining instances of the PRV. Multiple query randvars mean a finer granularity in the model after shattering, leading to more operations during LVE.

To compute an answer to a conjunctive query $\mathbf{Q}$, we shatter $G^\mathbf{Q}$ on $\mathbf{Q}$. We use LVE to compute a joint probability for $\mathbf{Q}$ and normalise. LJT still works for a singleton query of one randvar Q as we find a node k that covers Q, extract a model, namely the local model G_k and all messages to k, and perform LVE.

Example 9. After passing messages, LJT can answer, e.g., the queries $P(Res(eve),$ $Pub(eve, p_1))$ and $P(AttCnf(eve))$. For $\mathbf{Q}_1 = \{Res(eve), Pub(eve, p_1)\}$, nodes 2 and 3 cover the query randvars. The extracted model $G^{\mathbf{Q}_1}$ consists of G_2, G_3, and m_{12}. Shattering $G^{\mathbf{Q}_1}$ w.r.t. $\mathbf{Q}_1$ leads to five parfactors. LJT sums out $Pub(X, P)$, $X \neq eve$ and $P \neq p_1$ from the g_3 duplicate where X and P are not equal to eve and *article*, resulting in a parfactor g' with arguments $HoTpc$ and $AttCnf(X)$, $X \neq eve$. Next, it sums out $Res(X)$, $X \neq eve$, from the g_2 duplicate without eve, resulting in a parfactor g'' with arguments $HoTpc$ and $AttCnf(X)$, $X \neq eve$. Summing out $AttCnf(X)$, $X \neq eve$, from the product of g' and g'' yields a parfactor g''' with argument $HoTpc$. Summing out $AttCnf(eve)$ from the product of g_2 and g_3 where $X = eve$ and $P = p_1$ yields a parfactor $\hat{g}$ with arguments $HoTpc$, $Res(eve)$, and $Pub(eve, p_1)$. Last, LJT multiplies m_{12}, g''', and $\hat{g}$, sums out $HoTpc$, and normalises, leading to the queried distribution.

For $\mathbf{Q}_2 = \{AttCnf(eve)\}$, LJT can use node 2. It sums out $Res(X)$, $HoTpc$, and $AttCnf(X)$ where $X \neq eve$ from $G_2 \cup \{m_{12}, m_{32}\}$ after shattering.

Regarding the extensions to LJT, query answering changes substantially with *conjunctive queries* since the models for answering a query first need to be compiled as just described. *Counting* affects query answering in the sense that extracted models may contain PCRVs. LVE for query answering in LJT uses count conversion as well and can sum out PCRVs. Prior to a short empirical evaluation, we look at the extended LJT from a more theoretical viewpoint.

5   Theoretical Analysis

We look at soundness and best and worst case scenarios of LJT extended with counting and conjunctive queries.

Soundness. For the soundness of our LJT version, we assume that the original LJT and PCRVs and their handling in LVE and FO dtrees are sound. We first look at LJT with counting and then at LJT for conjunctive queries.

Theorem 1. *LJT with counting is sound, i.e., is equivalent to inference using a ground inference algorithm.*

Proof sketch. To compute answers to queries at a node with information present through a PP scheme, Shenoy and Shafer present three axioms for the operations marginalisation and combination on potential functions in a jtree [22]. Our definition of potential functions and PP scheme coincide with [22] with lifted summing out and lifted multiplication in the roles of marginalisation and combination fulfilling the axioms for local computations. The original LJT constructs a valid jtree in the form of an FO jtree with nodes containing randvars and jtree properties fulfilled.

The extended LJT still constructs a valid FO jtree as FO jtree construction is unchanged. PCRVs are handled during FO dtree construction and appear in parclusters accordingly. Since we assume a sound LVE with sound handling of count conversions and PCRVs, lifted summing out and lifted multiplication remain sound. Thus, message passing is still allowed and produces sound results: With sound LVE operations and a valid FO jtree, the algorithm carries out sound computations at the local models, sending sound information from one node to another. The same holds for query answering: With sound information at the nodes, LJT computes a correct answer for a query.

Next, we look at LJT with counting and conjunctive queries.

Theorem 2. *LJT with conjunctive queries is sound, i.e., is equivalent to inference using a ground inference algorithm.*

Proof sketch. Given that LJT with counting is sound, we have sound information at the nodes in the FO jtree. By way of constructing the model for the query randvars in a query, we combine all necessary information without duplicates as argued above. Given that LVE is sound, LJT computes a correct answer for a query on the extracted submodel.

Next, we look at best and worst case scenarios for LJT including what characteristics influence LJT runtimes.

Best and Worst Case Scenario. The extended LJT allows for efficient query answering given multiple queries. It imposes some static overhead due to FO jtree construction and message passing. After these steps, it answers queries based on typically smaller models compared to the input model G. If G changes, LJT ha to construct a new FO jtree.

Characteristics that influence runtimes include (i) during construction, the number of logvars and parfactors in G, (ii) during message passing, the number of nodes in the FO jtree, the size of the parclusters, and the degree of each node, and (iii) during query answering, the size of the model used for a query and the effort spent on building the model. The goal is to have efficient query answering with smallest models possible and spend effort on construction and message passing only once per input model.

In a worst case scenario (the same holds for LVE), LJT needs to ground all logvars in the model and perform inference at a propositional level to calculate correct results. In such a case, it cannot avoid groundings. Unfortunately, LJT may induce unnecessary groundings during message passing because calculating a message over PRVs with logvars may inhibit a reasonable elimination order. Simplified, the logvars of a PRV to eliminate need to be a superset of the logvars in affected separator PRVs. A separator PRV with the most logvars of all PRVs in a parcluster automatically results in at least a counting conversion and, in the worst case, groundings. Since messages are part of further calculations, groundings might carry forward. The results are still correct, but the LJT run degrades to a propositional algorithm run. For a detailed discussion, see [4].

For singleton queries, we gain the most if the model permits an FO jtree with few PRVs per parcluster. With a clever access function, LJT quickly identifies a parcluster for the query and sum out the few non-query PRVs. For conjunctive queries, the best case is if the nodes that cover all query PRVs are adjacent and form a submodel with few PRVs to eliminate. Needing the whole tree represents the worst case as LJT builds a submodel equal to the original model and do standard LVE, adding overhead without payoff. Over many queries, LJT offsets queries requiring a large model with queries using a small model.

After these theoretical considerations, we look at an empirical evaluation for our running example G_{ex}.

6   Empirical Evaluation

We have implemented a prototype of LJT with our extensions, named `exfojt` in this section. Taghipour provides a baseline implementation of GC-FOVE including its operators (available at https://dtai.cs.kuleuven.be/software/gcfove), named `gcfove`, which we use to test our implementation against. We use the `gcfove` operators in `exfojt`. We also implemented a propositional junction tree algorithm, named `jt`, as a reference point.

Standard lifting examples such as the smokers model are too simple, leading to an FO jtree with one node. Runtimes of `exfojt` on the standard examples compared to `gcfove` are slightly higher due to the static overhead for constructing an FO jtree (FO dtree construction, cluster calculation, parcluster conversion, merging). The resulting node carries the original model as a local model. Message passing does not apply with one node. Query answering takes the same time for each query as both carry out the same operations on the original model.

We use G_{ex} as input. We vary the domain sizes, yielding grounded model sizes $|gr(G_{ex})|$ between 3 and 241,000. We query each PRV once with one grounding, resulting in 6 queries,

- $HoTpc$,
- $Biz(m_1)$,
- $App(a_1)$,
- $Res(x_1)$,
- $AttCnf(x_1)$, and
- $Pub(x_1, p_1)$.

Which grounding we use is irrelevant for the calculations and the answer for the algorithms given a current model size, since the instances are interchangeable.

We compare runtimes for inference accumulated over the given queries, averaged over several runs. exfojt constructs an FO jtree comparable to the FO jtree in Fig. 3 with three nodes and passes messages (four messages). Then, it answers the given queries based on the local models and messages. jt follows the same protocol with propositional data structures and VE operations. The propositional jtrees have an increasing number of nodes with the largest cluster containing four randvars. While message passing takes longer in jtrees, QA is faster than QA in FO jtrees since only up to three randvars need to be eliminated without any accounting for logvars necessary. gcfove eliminates all non-query randvars from G_{ex} for each query. We do not compare against the original LJT version since our example model leads to groundings without counting. Its runtimes come close to the runtimes of jt.

Figure 4 shows runtimes for inference in milliseconds with $|gr(G_{ex})|$ on the x-axis, ranging from 3 to 241,000, both on log scale. The squares mark the runtimes for gcfove, the circles the runtimes for exfojt, and the filled triangles the runtimes for jt. With small models, jt outperforms both lifted approaches. With an increase of $|gr(G_{ex})|$, memory and time requirements of jt surge.

exfojt outperforms gcfove on all grounded model sizes, needing 43% to 51% of the time gcfove requires. The savings in runtime are mirrored in the number of LVE operations performed, with a maximum of 63 by gcfove versus 46 by exfojt. exfojt trades off runtime with storage, needing slightly more memory to store its FO jtree and messages at each node.

Since exfojt has some static overhead, we look at what point exfojt outperforms gcfove. Figure 5 shows runtimes on log scale accumulated over the six queries for $|gr(G_{ex})|$ = 102,050. The shape of the curves is identical over the different groundings with higher or lower runtimes. We ordered the given queries by increasing runtimes for gcfove. With the second query, gcfove needs marginally more time. With each passing query, exfojt saves more time compared to gcfove as it is able to answer queries based on one node.

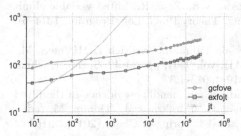
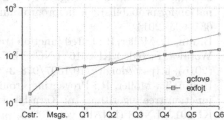

Fig. 4. Runtimes [ms] with $|gr(G_{ex})|$ ranging from 3 to 241,000 on log scales accumulated over 6 queries

Fig. 5. Runtimes [ms] on log scale with $|gr(G_{ex})|$ = 102,050 accumulating over 6 queries

Conjunctive queries that `exfojt` answers using one parcluster have similar runtimes compared to the singleton queries from above. With more complex queries that require more than one parcluster, runtimes increase since subtree identification takes longer and the models become larger. We do not compare runtimes for conjunctive queries as `gcfove` only supports singleton queries.

In summary, even in our small example model and only a prototype implementation, spending effort on an FO jtree pays off. LJT has even more potential when considering scenarios where the FO jtree structure remains the same and only parts of a model or other prior information changes.

7   Conclusion

We present extensions to LJT to answer multiple queries efficiently in the presence of symmetries in a model. We formally specify the different steps of LJT and incorporate the lifting tool of counting to lift computations where LJT previously needed to ground. We extend the scope of LJT by allowing conjunctive queries and handling them efficiently. These extensions provide us with a deeper understanding of how LVE and FO jtrees interact. If a model allows for a lifted run, i.e., without groundings, we speed up runtimes significantly for answering multiple queries compared to the original LJT and GC-FOVE.

We currently work on adapting LJT to incrementally changing models. Other interesting algorithm features include parallelisation, construction using hypergraph partitioning, and different message passing strategies as well as using local symmetries. Additionally, we look into areas of application to see its performance on real-life scenarios.

References

1. Ahmadi, B., Kersting, K., Mladenov, M., Natarajan, S.: Exploiting symmetries for scaling loopy belief propagation and relational training. Mach. Learn. **92**(1), 91–132 (2013)
2. Bellodi, E., Lamma, E., Riguzzi, F., Costa, V.S., Zese, R.: Lifted variable elimination for probabilistic logic programming. Theory Pract. Log. Program. **14**(4–5), 681–695 (2014)
3. Braun, T., Möller, R.: Lifted junction tree algorithm. In: Friedrich, G., Helmert, M., Wotawa, F. (eds.) KI 2016. LNCS (LNAI), vol. 9904, pp. 30–42. Springer, Cham (2016). https://doi.org/10.1007/978-3-319-46073-4_3
4. Braun, T., Möller, R.: Preventing groundings and handling evidence in the lifted junction tree algorithm. In: Kern-Isberner, G., Fürnkranz, J., Thimm, M. (eds.) KI 2017. LNCS (LNAI), vol. 10505, pp. 85–98. Springer, Cham (2017). https://doi.org/10.1007/978-3-319-67190-1_7
5. van den Broeck, G.: Lifted inference and learning in statistical relational models. Ph.D. thesis, KU Leuven (2013)

6. van den Broeck, G., Niepert, M.: Lifted probabilistic inference for asymmetric graphical models. In: AAAI 2015 Proceedings of the 29th Conference on Artificial Intelligence, pp. 3599–3605 (2015)

7. Chavira, M., Darwiche, A.: Compiling Bayesian networks using variable elimination. In: IJCAI 2007 Proceedings of the 20th International Joint Conference on Artificial Intelligence, pp. 2443–2449 (2007)

8. Choi, J., Amir, E., Hill, D.J.: Lifted inference for relational continuous models. In: UAI 2010 Proceedings of the 26th Conference on Uncertainty in Artificial Intelligence, pp. 13–18 (2010)

9. Darwiche, A.: Recursive conditioning. Artif. Intell. **2**(1–2), 4–51 (2001)

10. Darwiche, A.: Modeling and Reasoning with Bayesian Networks. Cambridge University Press, Cambridge (2009)

11. Das, M., Wu, Y., Khot, T., Kersting, K., Natarajan, S.: Scaling lifted probabilistic inference and learning via graph databases. In: Proceedings of the SIAM International Conference on Data Mining, pp. 738–746 (2016)

12. Gogate, V., Domingos, P.: Exploiting logical structure in lifted probabilistic inference. In: Working Note of the Workshop on Statistical Relational Artificial Intelligence at the 24th Conference on Artificial Intelligence, pp. 19–25 (2010)

13. Gogate, V., Domingos, P.: Probabilistic theorem proving. In: UAI 2011 Proceedings of the 27th Conference on Uncertainty in Artificial Intelligence, pp. 256–265 (2011)

14. Jensen, F.V., Lauritzen, S.L., Olesen, K.G.: Bayesian updating in recursive graphical models by local computations. Comput. Stat. Q. **4**, 269–282 (1990)

15. Kersting, K., Ahmadi, B., Natarajan, S.: Counting belief propagation. In: UAI 2009 Proceedings of the 25th Conference on Uncertainty in Artificial Intelligence, pp. 277–284 (2009)

16. Koller, D., Friedman, N.: Probabilistic Graphical Models: Principles and Techniques. The MIT Press, Cambridge (2009)

17. Lauritzen, S.L., Spiegelhalter, D.J.: Local computations with probabilities on graphical structures and their application to expert systems. J. Royal Stat. Soc. Ser. B Methodol. **50**, 157–224 (1988)

18. Milch, B., Zettelmeyer, L.S., Kersting, K., Haimes, M., Kaelbling, L.P.: Lifted probabilistic inference with counting formulas. In: AAAI 2008 Proceedings of the 23rd Conference on Artificial Intelligence, pp. 1062–1068 (2008)

19. Poole, D., Zhang, N.L.: Exploiting contextual independence in probabilistic inference. J. Artif. Intell. **18**, 263–313 (2003)

20. de Salvo Braz, R., Amir, E., Roth, D.: Lifted first-order probabilistic inference. In: IJCAI 2005 Proceedings of the 19th International Joint Conference on Artificial Intelligence (2005)

21. Shafer, G.R., Shenoy, P.P.: Probability propagation. Ann. Math. Artif. Intell. **2**(1), 327–351 (1990)

22. Shenoy, P.P., Shafer, G.R.: Axioms for probability and belief-function propagation. Uncertain. Artif. Intell. **4**(9), 169–198 (1990)

23. Singla, P., Domingos, P.: Lifted first-order belief propagation. In: AAAI 2008 Proceedings of the 23rd Conference on Artificial Intelligence, pp. 1094–1099 (2008)

24. Taghipour, N.: Lifted probabilistic inference by variable elimination. Ph.D. thesis, KU Leuven (2013)

25. Taghipour, N., Davis, J., Blockeel, H.: First-order decomposition trees. In: Advances in Neural Information Processing Systems 26, pp. 1052–1060. Curran Associates, Inc. (2013)

26. Taghipour, N., Fierens, D., Davis, J., Blockeel, H.: Lifted variable elimination: decoupling the operators from the constraint language. J. Artif. Intell. Res. **47**(1), 393–439 (2013)
27. Vlasselaer, J., Meert, W., van den Broeck, G., Raedt, L.D.: Exploiting local and repeated structure in dynamic Baysian networks. Artif. Intell. **232**, 43–53 (2016)
28. Zhang, N.L., Poole, D.: A simple approach to Bayesian network computations. In: Proceedings of the 10th Canadian Conference on Artificial Intelligence, pp. 171–178 (1994)

Representing and Reasoning About Logical Network Topologies

Shaun Voigt, Catherine Howard[⊠], Dean Philp,
and Christopher Penny

Defence Science and Technology Group, Edinburgh, Adelaide, Australia
{shaun.voigt,catherine.howard,
dean.philp}@dst.defence.gov.au

Abstract. For network analysts, constructing a representation, and developing an understanding, of logical network topologies is crucial for a wide range of cyber security applications. However, constructing a representation of logical network topologies is difficult. This paper presents three novel ontologies; the Internet Protocol (IP) Ontology, the Open Shortest Path First (OSPF) Ontology and the Border Gateway Protocol (BGP) Ontology. These ontologies provide a common, technology independent syntax and semantics for complex communication network concepts. The semantic and syntactic interoperability provided by these ontologies enables data from disparate, heterogeneous sources, such as network diagrams, router configuration files and routing protocol messages, to be consistently represented, which facilitates information fusion. The approach presented in this paper allows domain knowledge to be encoded in an intuitive manner, facilitates knowledge discovery by automated reasoning, and facilitates the process of making specialist knowledge and tradecraft accessible to non-expert network analysts.

Keywords: Ontologies · Network data · Network topologies

1 Introduction

For network analysts, constructing a representation, and developing an understanding, of logical network topologies[1] is crucial for a wide range of cyber security applications such as traffic path estimation, network monitoring and management [1], network vulnerability assessment and defence [2], identifying network boundaries and understanding the propagation of BGP hijacks. However, constructing a representation of logical network topologies is difficult, especially at Internet scale. The Internet is the largest, most complex artificially deployed system in existence [3] and there are many disparate, heterogeneous sources of data which could potentially be used. The application of automated information fusion techniques, and the associated underlying

---

[1] The topology of a network is the arrangement of the various network elements, such as routers, computers and links, within the network. The topology of a network may be depicted physically or logically. The physical topology of a network is the arrangement of the physical components of the network, including the location of devices and cables. While the logical topology illustrates how information flows through the network.

© Crown 2018
M. Croitoru et al. (Eds.): GKR 2017, LNAI 10775, pp. 73–83, 2018.
https://doi.org/10.1007/978-3-319-78102-0_4

knowledge representation and automated reasoning techniques, could assist in addressing these scale and complexity issues. The fusion of data from multiple sources can provide a more detailed representation of the logical network topology than the representation provided by any individual data source in isolation. However, the automated fusion of data from disparate, heterogeneous sources requires semantic and syntactic interoperability. To provide this interoperability, this research adopted an ontological approach to knowledge representation.

There is a dearth of literature on the use of ontologies for constructing representations of logical communication network topologies; ontologies have been developed for network planning and design (e.g., [4]), network measurement and monitoring (e.g., [5]) and the provisioning, configuration and management of virtual or physical network resources (e.g., [6–13]). However, none of these ontologies provided a formal specification of the IP, OSPF and BGP concepts required by this research, at the required level of detail. Hence this research developed three novel ontologies which can be used to represent complex communication network concepts; the Internet Protocol (IP) Ontology, the Open Shortest Path First (OSPF) Ontology and the Border Gateway Protocol (BGP) Ontology. This paper presents these three novel ontologies.

The rest of this paper is structured as follows. Section 2 describes the data sources utilised by this research. Section 3 presents the three novel ontologies, justifies the selection of the Web Ontology Language (OWL) as the implementation language and describes the knowledge representation and reasoning processes. Section 4 provides an example of producing a representation of a logical network topology using some of the data sources described in Sect. 2 and the ontologies and processes outlined in Sect. 3. Section 5 presents a brief discussion while Sect. 6 presents the conclusions.

2 The Data Sources

There are many disparate, heterogeneous sources of network data which could potentially be used to construct representations of logical network topologies. This research focused on being able to represent, fuse and reason about six such sources; network diagrams, router configuration files, routing tables, Open Shortest Path First (OSPF) Link State Advertisements (LSAs), Border Gateway Protocol (BGP) update messages and open source data.

A network diagram is a visual representation of the physical or logical topology of a network. It depicts the nodes (including routers, switches, servers, printers and hosts) in the network and the connections between them.

A router's configuration file contains all the commands required to configure the router. It contains information such as the IP addresses of the router's interfaces, the routing protocols used on each interface and the metrics used by link state routing protocols[2].

For each reachable destination, a routing table lists the network element which is next along the path to the destination. When an IP packet arrives, a router uses this

---

[2] In a link state routing protocol, each router constructs a map of the connectivity of the network in which it resides.

table to determine the interface on which to forward the packet based on its destination IP address.

OSPF [14] is the most widely used interior gateway protocol[3] (IGP) on the Internet [15]. Link State Advertisements (LSAs) are the basic communication mechanism of OSPF. There are eleven different types of LSAs. This research utilises Router (also referred to as Type 1) and Network (also referred to as Type 2) LSAs. A Router LSA contains information about all routers and networks which are directly connected to the originating router. A Network LSA includes the network identifier, subnet mask and a list of routers which are joined together by the broadcast domain[4].

BGP [16] is an exterior gateway protocol; it is used to facilitate inter Autonomous System[5] (AS) relationships by exchanging routing and reachability information among ASes on the Internet. When a BGP session is initialised between routers, update messages are sent to exchange routing information until the complete BGP routing table has been exchanged. A router advertises the networks which are reachable via each of its neighbours and how many hops away each network is.

There is a myriad of open source information which could potentially be useful. This research focused on utilising some of the data available from the Center for Applied Internet Data Analysis (CAIDA)[6] [17], including the:

- AS name, number and owner;
- Networks that an AS is the registered owner of;
- Networks advertised by an AS; and
- Inter-AS relationships that an AS participates in (i.e., peering and customer-provider relationships).

From the above descriptions, it can be seen that the six data types are disparate and heterogeneous. The data itself is complex, relational data, which is not easily understood by analysts without specialist communication network knowledge and experience.

3   The Ontologies and the Knowledge Representation and Reasoning Process

The Web Ontology Language (OWL) [18] was selected to implement the three ontologies because, among other reasons:

- OWL is explicitly designed to support the integration of data from multiple sources [19].

---

[3] Interior gateway protocols manage the routing of traffic within individual ASes.

[4] A broadcast domain is a logical division of a network, in which all devices can reach each other by broadcast at the data link layer. For example, a multi-access network is a single broadcast domain. Ethernet is also an example of a broadcast domain.

[5] An Autonomous System is a network, or collection of networks, which are managed or supervised by a single administrative entity or organisation.

[6] CAIDA is a collaboration of government, research and commercial entities aimed at promoting greater cooperation in the engineering and maintenance of the global Internet infrastructure.

- OWL and RDF are well suited to the representation of complex, relational data [19].
- RDF triples can be represented as semantic networks [20], which are a natural representational match for logical network topologies (which can be represented as undirected graphs [19]).
- OWL provides an explicit separation between syntax and semantics.
- OWL can be coupled with semantic reasoners and rule-based languages, such as the Semantic Web Rule Language (SWRL) [18], to support automated reasoning.
- OWL allows ontologies to reuse classes and properties from existing, published ontologies [12, 19].

In this research, OWL and RDF were used during the knowledge representation process and SWRL and the SPARQL Protocol and RDF Query Language (SPARQL) [18] were used during the reasoning process.

Figure 1 shows the hierarchy of the IP, OSPF and BGP ontologies, with the OSPF and BGP ontologies inheriting classes, data properties and object properties from the IP ontology. Because there is insufficient room to present the full OWL functional syntax, the ontologies will be presented using relational diagrams. Relational diagrams depict the set of classes, data properties and object properties in an ontology. In relational diagrams, classes are represented by large rectangles, with the name of the class in bold print, and object properties are underlined and linked to their range types by directed lines.

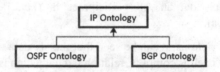

Fig. 1. The inheritance hierarchy of the ontologies.

The IP Ontology, shown in Fig. 2, represents concepts at the IP layer (i.e., Layer 3 of the OSI model [21]). As the IP ontology focuses on the IP layer, Layer 2 devices (such as switches) and physical connections (such as cables) are not included. The set of classes in the IP ontology is **C** = {*Network Element, Network, Router, Computer, Interface, Route Entry, Default Route Entry, Directly Connected Route Entry*}.

The OSPF ontology, shown in Fig. 3, extends the IP Ontology by introducing OSPF specific concepts such as OSPF areas[7] and Area Border Routers (ABRs)[8]. The set of classes in the ontology is **C** = {*Network Element, Network, Router, Computer, Interface, Area, Route Entry, Default Route Entry, Directly Connected Route Entry, OSPF Summary Route Entry*}. From Fig. 3 it can be seen that the OSPF Ontology inherits classes from the IP ontology (e.g., the *Network Element, Network* and *Router* classes), specialises some of the object properties of these classes (e.g., the *hasNeighbour* object property of the *Router* class has a new *hasOSPFRouterNeighbour*

---

[7] An OSPF network can be subdivided into multiple routing areas in order to simplify administration or optimise traffic flow or resource utilisation.

[8] ABRs are routers which have interfaces in multiple areas.

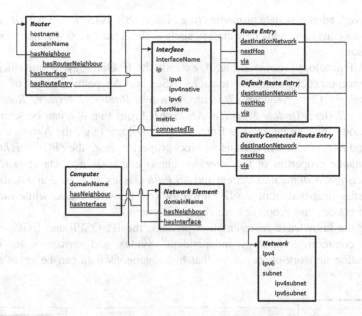

Fig. 2. The relational diagram of the IP ontology.

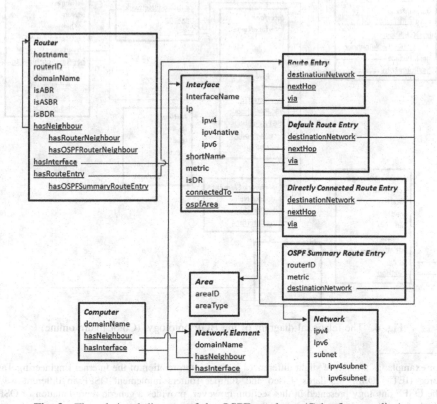

Fig. 3. The relational diagram of the OSPF ontology. (Color figure online)

specialisation), adds new data properties (e.g., the *isABR*, *isASBR* and *isBDR* properties of the *Router* class) and adds new classes such as the *Area* and *OSPF Summary Route Entry* classes.

The BGP ontology, shown in Fig. 4, extends the IP Ontology by introducing BGP specific concepts, such as update messages, ASes and AS paths. The set of classes in the ontology is **C** = {*Network Element, Network, Router, Interface, Route Entry, Autonomous System, Update Message, AS Path*}. From Fig. 4 it can be seen that the BGP Ontology inherits the classes from the IP ontology (e.g., the *Network Element, Network* and *Router* classes), adds new object properties (e.g., the *eBGPNeighbour* and *iBGPNeighbour* properties of the *Interface* class) and adds new classes such as the *Update Message, Autonomous System* and *AS Path* classes. In Figs. 3 and 4, the classes and properties inherited from the IP Ontology are shown in black, while the new or specialised classes and properties are shown in green.

During the knowledge representation process, the IP, OSPF and BGP ontologies provide a common, technology independent[9] syntax and semantics for complex communication network concepts, so that heterogeneous data can be encoded into a

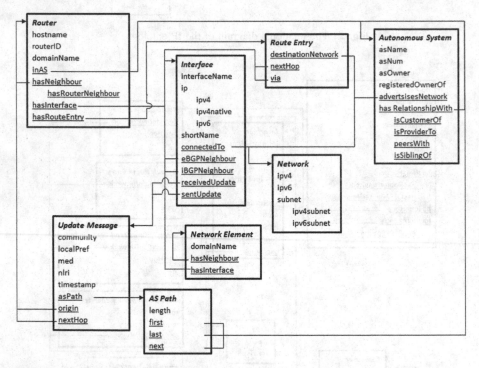

Fig. 4. The relational diagram of the BGP ontology. (Color figure online)

---

[9] For example, as a result of slight differences in their interpretation of the Internet Engineering Task Force (IETF) OSPF standards, Cisco and Juniper routers implement OSPF in different ways. The OSPF ontology presented in this section, however, provides a generic representation of OSPF which is not dependant on the specific implementation technology.

consistent representation. Once encoded, the data is in RDF triple format and is referred to as instance data. The instance data are stored in a triple store in the knowledge base.

Context specific rules enable subject matter experts (SMEs) to encode specialist knowledge or tradecraft using SWRL. Examples of context specific rules are provided in Sect. 4. Context specific rules can be used to perform data cleaning and information fusion. During the reasoning process, using the ontologies and context specific rules, the rule-based inference engine performs reasoning over the instance data in the knowledge base. The reasoning process is a forward-chaining, data-driven process, whereby new information can trigger the execution of additional context specific rules.

4 Fusion Example

Consider the scenario shown in Fig. 5. In this scenario, there are two ASes, *AS10143* and *AS1221*, which are connected by a single inter-AS relationship. *AS10143* has two routers *AS10143R1* and *AS10143R2*. *AS1221* has one router *AS1221R1*. *AS1221R1* has an external BGP relationship with *AS10143R1*. *AS10143* is using OSPF as its IGP. Suppose that the available information sources include:

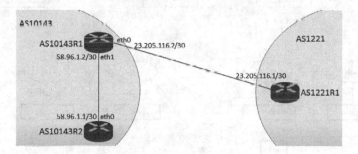

Fig. 5. The scenario under consideration.

- Open source CAIDA data pertaining to *AS10143* and *AS1221*;
- A BGP update message sent from *AS10143R1* to *AS1221R1;*
- Router configuration files for *AS10143R1* and *AS10143R2;* and
- OSPF Router LSAs issued by *AS10143R1* and *AS10143R2*.

However, for this example, it is assumed that no router configuration files, OSPF Router LSAs or BGP update messages are available for *AS1221*.

Using context specific rules such as:

$$If\ two\ Network\ objects\ have\ the\ same\ ipv4subnet\ value,$$
$$then\ the\ two\ Networks\ objects\ are\ the\ same\ object\ ; \tag{1}$$

$$If\ two\ Interface\ objects\ have\ the\ same\ ipv4\ value,$$
$$then\ the\ two\ Interface\ objects\ are\ the\ same\ object; \tag{2}$$

$$If\ two\ Router\ objects\ have\ the\ same\ routerID\ value,$$
$$then\ the\ two\ Router\ objects\ are\ the\ same\ object;\ and \tag{3}$$

$$If\ two\ AS\ objects\ have\ the\ same\ asNum,\ and\ it\ is\ a\ public\ asNum,$$
$$then\ it\ is\ the\ same\ AS, \tag{4}$$

the data from the aforementioned sources can be fused to produce the representation of the logical network topology shown in Fig. 6. This semantic network contains

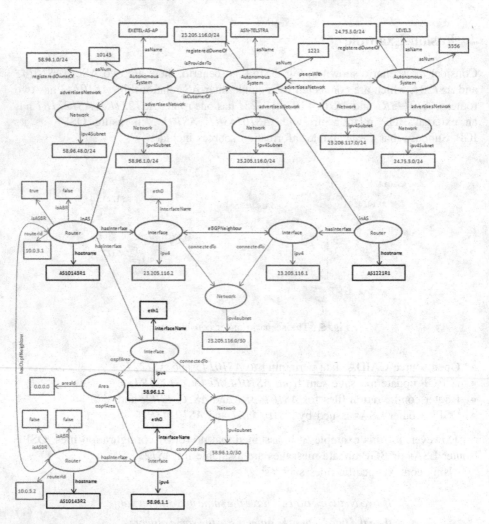

Fig. 6. The semantic network resulting from the fusion of all the available data. Blue represents open source CAIDA data, red represents the data obtained from the BGP update message sent from *AS10143R1* to *AS1221R1*, black represents the data obtained from *AS10143R1*'s and *AS10143R2*'s configuration files and green represents the data obtained from *AS10143R1*'s and *AS10143R2*'s LSAs. (Color figure online)

information about interfaces, routers, networks, areas and ASes and the relationships between these concepts. It combines intra-AS connectivity information provided by OSPF with inter-AS connectivity information provided by BGP and CAIDA, allowing an analyst to see the connection between Internet level routing and the private network infrastructure of EXETEL-AS-AP[10]. It can be seen that the enriched representation of the network's logical topology provided by Fig. 6 is more detailed and accurate than the representation provided by any individual data source in isolation.

5 Discussion

The IP, OSPF and BGP ontologies provide a consistent way to represent complex communication network concepts. The ontologies are easily extensible. They support communication and information sharing, automated reasoning and the reuse of domain knowledge. They also limit ambiguity and make domain assumptions explicit. Making the domain assumptions explicit makes it possible to change these assumptions if the knowledge about the domain changes. Being able to represent the resulting network topologies as semantic networks facilitates human understanding.

The three ontologies all contain concepts at a range of abstractions; from high level concepts such as ASes and networks through to low level concepts such as router interfaces. This allows:

- Information at different levels of abstraction to be represented and fused. For example, Fig. 6 depicts EXETEL-AS-AP in a high level of detail while ASN-TELSTRA, where less information is available, is represented at a more abstract level.
- Networks to be represented at different levels of abstraction. For example, using the same ontologies and same data sources, the same network could be represented by a semantic network containing:
  - Networks, routers, interfaces, IP addresses, subnets, interface names, host names and the *hasInterface* and *connectedTo* relationships; or
  - Routers, host names and the *hasRouterNeighbour* relationships.
- The same concepts to be used in different ways.

Being able to represent information at different levels of abstraction is important because abstraction allows:

- Complex data to be simplified. This simplification can reduce the complexity of both data analysis and visualisation and can enable complex data to be hidden from non-expert analysts.
- The application of graph theoretic techniques at an abstracted level, rather than the lowest level of detail, where the size and complexity of the semantic network may preclude their use.

---

[10] Synthetic data has been used for the private network infrastructure in order to demonstrate the fusion techniques.

The context specific rules discussed in Sect. 3 can provide a natural way for SMEs to encode their specialist knowledge or tradecraft, potentially making this knowledge more accessible to non-experts analysts. Because OWL is a declarative language, rules can be developed which work for a large number of instances. Rules can be generic, and hence applicable to all types of networks, or they can be specific for specific types of networks (for example, content distribution networks). The ability to use different context specific rule sets, based on the situation, provides a level of flexibility. However rules can have a number of limitations. For example, the quality of the rule base developed for a particular domain will be dependent on the experience and point of view of the SMEs who construct it, so there may be gaps, overlaps and inconsistencies. Encoding rules can also be difficult; knowledge elicitation is manual and can be error prone. It can be easy, for example, to create contradictory rules. A large rule set can be difficult to maintain and update.

6   Conclusions

This paper presented three novel ontologies; the IP Ontology, the OSPF Ontology and the BGP Ontology. These ontologies provide a common, technology independent syntax and semantics for complex communication network concepts. The semantic and syntactic interoperability provided by the three ontologies allows data from disparate, heterogeneous sources to be consistently represented, which facilitates information fusion.

The approach presented in this document allows domain knowledge to be encoded in an intuitive manner, facilitates knowledge discovery by automated reasoning, and facilitates the process of making specialist knowledge and tradecraft accessible to non-expert network analysts.

While ontological approaches to knowledge representation have many strengths, the quality of an ontology developed for a particular domain will always be dependent on the experience and point of view of the SMEs who build it, so there are always gaps, overlaps and inconsistencies. However, this is true of any knowledge representation technique.

Acknowledgements. Part of this work was conducted using the Protégé resource [22], which is supported by grant GM10331601 from the National Institute of General Medical Sciences of the United States National Institutes of Health.

References

1. van der Ham, J., Ghijsen, M., Grosso, P., de Laat, C.: Trends in Computer Network Modeling Towards the Future Internet. https://arxiv.org/pdf/1402.3951v2.pdf. Accessed Oct 2016
2. Motamedi, R., Rejaie, R., Willinger, W.: A survey of techniques for internet topology discovery. IEEE Commun. Surv. Tutor. **17**(2), 1044–1065 (2013)
3. Ioannou, P.A., Pitsillides, A.: Modeling and Control of Complex Systems. CRC Press, Boca Raton (2008)

4. Rahman, M., Pakstas, A., Wang, F.Z.: Towards communications network modelling ontology for designers and researchers. In: Proceedings of the International Conference on Intelligent Engineering Systems, London, England (2006)
5. MOMENT - Monitoring and Measurement in the Next Generation Technologies. http://www.salzburgresearch.at/en/projekt/moment_en/. Accessed Oct 2016
6. Yeung, D., Qu, Y., Zhang, J., Chen, I., Lindem, A.: Yang Data Model for OSPF Protocol. https://tools.ietf.org/html/draft-ietf-ospf-yang-01. Accessed Oct 2016
7. Zhdankin, A., Patel, K., Clemm, A., Hares, S., Jethanandani, M., Liu, X.: Yang Data Model for BGP Protocol. https://tools.ietf.org/html/draft-zhdankin-idr-bgp-cfg-00. Accessed Oct 2016
8. Common Information Model. http://www.dmtf.org/standards/cim. Accessed Aug 2015
9. Strassner, J.: DEN-ng: achieving business-driven network management. In: Proceedings of the IEEE/IFIP Network Operations and Management Symposium (2002)
10. van der Ham, J., Dijkstra, F., Lapacz, R., Brown, A.: The network markup language; a standardized network topology abstraction for inter-domain and cross-layer network applications. In: Proceedings of the TERENA Networking Conference, Maastricht, Netherlands (2013)
11. van der Ham, J., Dijkstra, F., Travostino, F., Andree, H., de Laat, C.: Using RDF to describe networks. Future Gener. Comput. Syst. 22(8), 862–867 (2006)
12. Ghijsen, M., van der Ham, J., Grosso, P., Dumitru, C., Zhu, H., Zhao, Z., de Laat, C.: A semantic-web approach for modelling computing infrastructures. J. Comput. Electr. Eng. 39, 2553–2565 (2013)
13. Network Innovation over Virtualized Infrastructures. http://www.fp7-novi.eu/index.php. Accessed Oct 2016
14. Moy, J.: RFC 2328 - OSPF Version 2. https://www.ietf.org/rfc/rfc2328.txt. Accessed Oct 2016
15. Nakibly, G., Gonikman, D., Kirshon, A., Boneh, D.: Persistent OSPF attacks. In: Proceedings of the Nineteenth Annual Network and Distributed System Security Conference (2012)
16. Rekhter, Y., Li, T., Hares, S.: RFC 4271 - A Border Gateway Protocol 4 (BGP-4). https://www.ietf.org/rfc/rfc4271.txt. Accessed Oct 2016
17. Center for Applied Internet Data Analysis. www.caida.org. Accessed Oct 2016
18. Antoniou, G., van Harmelen, F.: A Semantic Web Primer. MIT Press, Cambridge (2004)
19. Reynolds, D., Thompson, C., Mukerji, J., Coleman, D.: An Assessment of RDF/OWL Modelling. Digital Media Systems Laboratory, HP Laboratories Bristol, HPL-2005-189 (2005)
20. Sowa, J.: Semantic networks. In: The Encyclopedia of Artificial Intelligence, 2nd edn. (1987)
21. OSI Model. https://en.wikipedia.org/wiki/OSI_model. Accessed Oct 2016
22. Protege. http://protege.standford.edu/. Accessed Oct 2016

From Enterprise Concepts to Formal Concepts: A University Case Study

Jamie Caine(iD) and Simon Polovina(✉)(iD)

Conceptual Structures Research Group, Department of Computing,
Communication and Computing Research Centre,
Sheffield Hallam University, Sheffield, UK
{j.caine,s.polovina}@shu.ac.uk

Abstract. A business enterprise is more than its buildings, equipment or financial statements. Enterprise Architecture frameworks thus include a metamodel that attempts to bring together all the enterprise concepts including the visible entities into a unified conceptual structure. Using a case study based upon the institution of the authors, the effectiveness of this conceptual structure is explored in two fold. Firstly, a simple example using familiar concepts such as the physical location of the authors' institution. Secondly, a more detailed example that includes the key enterprise concepts that currently exist within that institution. The metamodel is stated in Conceptual Graphs then mapped from these graphs' triples into transitive Formal Concept binaries using the *CGFCA* software. Misalignments within the enterprise concepts discovered from the derived formal concepts are highlighted in both case examples, hence pointing towards the wider applicability of this approach.

1 Introduction

A business enterprise is more than just the sum of its buildings, equipment or financial statements. Such visible entities are simply the structures that follow from its strategy, which is just as real. Strategy is moreover the driving entity, without which the enterprise falters. Like many other disciplines, business modelling practitioners (such as enterprise architects) rely on useful conceptual models that underpin enterprise activity. The underlying enterprise concepts in these models capture the purpose of the enterprise (why it exists) and articulated through its strategy. To achieve its strategic goals, the enterprise concepts extend into the enterprise's lower level tactical and operational goals that include its locations, finance, assets (e.g. buildings, trading stock, information technology), staff and an organisational structure. History however continues to show these entities becoming the drivers resulting in the emergence of bureaucratic structures, inter-departmental conflicts, inadequate computer systems and other experiences where we have 'The tail wagging the dog' i.e. strategy is lost and ends up following structure [3]. Put another way, the operational enterprise concepts overtake the strategic enterprise concepts when it should be the other way round.

© Springer International Publishing AG, part of Springer Nature 2018
M. Croitoru et al. (Eds.): GKR 2017, LNAI 10775, pp. 84–96, 2018.
https://doi.org/10.1007/978-3-319-78102-0_5

To address this phenomenon the paper is structured as follows. Enterprise concepts are introduced and discussed through the notion of enterprise architecture and the formal depiction of the enterprise concepts through ontology, semantics and metamodels. The relevance and use of Conceptual Structures is then addressed by illustrating two examples of the same case study, Sheffield Hallam University (SHU) being the institution of the authors. The first example is a simplified example with reference to a simple structured metamodel. The second example reflects a more accurate depiction of the concepts and transitive aspects that embody SHU's strategy [16]. This entails a decomposition of the SHU strategy by starting with an uppermost concept 'Forces and Trends' that influence strategy and ends in Process Performance Indicator (PPI). This section also explicates Conceptual Graphs (CGs), Formal Concept Analysis (FCA) and the *CGFCA* software and how they are used. For both examples, FCA Lattices generated from the CGs are iterated to correct the model (and metamodel in the simplified SHU example). It is through the corrections that we further understand the value that formal concepts bring to enterprise concepts. This is followed by a discussion of the further significance of this work, culminating in the paper's conclusions.

2 Enterprise Architecture

Enterprise Architecture (EA) recognises that enterprises are best understood by a holistic approach that explicitly refers to every important issue from every important perspective [20]. Hence all the enterprise concepts need to relate to each other.

2.1 Ontology, Semantics and Metamodels

EA arose from Zachman's original Information Systems Architecture Framework [12,20]. Zachman's EA framework places the enterprise concepts in cells that are interrelated through a simple two-dimensional matrix, consequently referred as an enterprise ontology [14,19]. The Open Group Architecture Framework (TOGAF) articulates the semantics in such an ontology by formally defining the relations between the enterprise concepts (entities) in a content metamodel rather than simply relying on their position in a matrix (or table) like Zachman [4,5]. A metamodel is the model about the model. The TOGAF metamodel formally describes the model to which every enterprise conforms, thereby embodying enterprise concepts. The EA metamodels have been comprehensively enhanced by the enterprise standards body LEADing Practice in association with the Global University Alliance [1,11].

3 Conceptual Structures

In his seminal text, Sowa describes Conceptual Structures (CS) as "Information Processing in Mind and Machine" [15]. Enterprises essentially arise as acts of human creativity in identifying business opportunities or other organisational solutions to social needs (e.g. government bodies, charities, schools or universities to name a few). Formal depictions of the metamodels (and the models that they in turn represent) enable them to be computable. Software tools potentially bring the productivity of computers to bear on interpreting the enterprise concepts, offering more expressive knowledge-bases leading to better decision-making. CS brings human creativity and computer productivity into the same mindset; CS thus offers an attractive proposition for capturing, interrelating and reasoning with enterprise concepts.

3.1   A Simplified Case Study of SHU

To clarify the approach, and explore the value of CS to enterprise concepts, a simple case study is now presented. For ease of understanding a much-simplified metamodel is used as well as a simplified description of the case study, which is Sheffield Hallam University (SHU) where the author of this paper is employed. SHU is a large public university located in Sheffield in the UK. Remembering that the term enterprise does not only apply to profit-making businesses, SHU's strategy is epitomised by the term 'Transforming Lives'. SHU meets this strategy through its location in Sheffield and the staff it employs (noting that these aspects are chosen from all its visible entities for simplicity.) The success of its strategy as realised through its staff and location (in this simplified example) is measured by Key Performance Indicators (KPIs). One such KPI in the UK is the National Student Survey ('the NSS', www.thestudentsurvey.com).

3.2   Conceptual Graphs

To demonstrate CS, Sowa devised Conceptual Graphs (CGs). CGs are essentially a system of logic that express meaning in a form that is logically precise, humanly readable, and computationally tractable. CGs serve as an intermediate language for translating between computer-oriented formalisms and natural languages. CGs graphical representation serve as a readable, but formal design and specification language [7,13].

Figure 1 reveals that CGs follow an elementary concept→relation→concept structure, which describes the ontology and semantics of the enterprise concepts as explained earlier. The CGs are thus directed graphs that capture the metamodel at the logical level including its direction of flow. Figure 1's left-hand side CG is the metamodel for our simple example, and the right-hand side is the specialised model for SHU that conforms to the metamodel. The type label Vision & Mission, Enterprise, Place, and KPI are each specialised by gaining a defined referent, which is an instantiation of the type label. The referent is 2020-Strategy.docx (a written document), Sheffield Hallam University (the enterprise), Sheffield (SHU's location) and {NSS-data...} (a structured

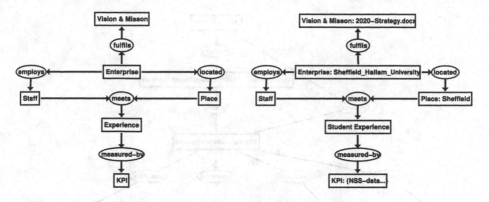

Fig. 1. Metamodel and SHU, in CGs

data source) for each type label respectively[1]. The type label `Experience` was specialised to `Student Experience`, which is `Experience`'s subtype.

3.3   An Expanded Example of the SHU Case Study

The expanded example depicts SHU's 'Transforming Lives' strategy and the distinctive four strategic pillars that it encompasses [16]. Due to becoming too large by being represented as one large CG (Conceptual Graph), the CGs for this example are shown by four modularised CGs i.e. Figs. 2, 3, 4 and 5.

These modified CGs have duplicate referents that are hence co-referent. The CGs can thereby be rejoined through the CGs join operation from their co-referent links [13]. These CGs draw upon the LEADing Practice Strategy Metamodel reference content [11,17].

Traditional strategy formulation accommodates the impact that forces and trends can have on organisational strategy [10]. Given this more accurately describes SHU's Enterprise Architecture, the concept of forces and trends are included within this model. Each strategic pillar is realised through goals and objectives that are each then measured by a Key Performance Indicator (KPI) that is current to SHU's strategy. Each KPI then measures a function followed by the role performing the function that in turn delivers a service. The model culminates in the Process Performance Indicator (PPI) concept that addresses each process deriving out of each strategic pillar in one PPI concept.

---

[1] { } denote 'plural' referents, meaning they hold more than one referent. Here NSS-data may be one of many datasets that collectively provide KPIs of SHU's strategy and shown simply to illustrate multiple cardinality of concepts. The Staff type label would also have a plural referent to depict the many staff that SHU employs. Plural referents are however not elaborated further for this simple case study's purposes.

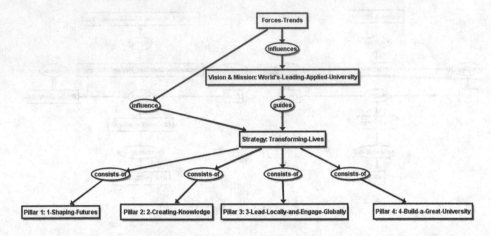

Fig. 2. Modified SHU part 1 of 4, in CGs

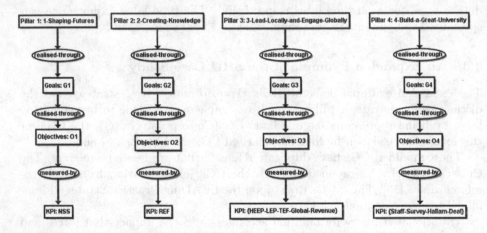

Fig. 3. Modified SHU part 2 of 4, in CGs

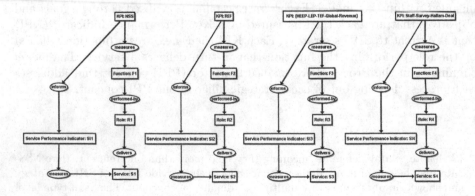

Fig. 4. Modified SHU part 3 of 4, in CGs

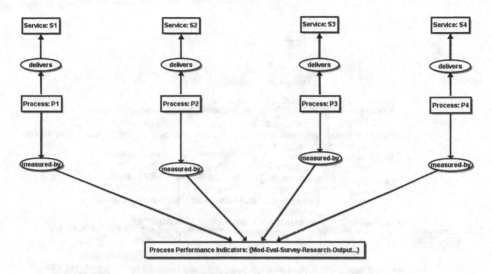

Fig. 5. Modified SHU part 4 of 4, in CGs

3.4 Formal Concept Analysis

Formal Concept Analysis (FCA) adds a mathematical level to the logical level captured in CGs [6]. The FCA formal context is generated from the CGs by the *CGFCA* software[2] [2]. Essentially, this software transforms CGs' underlying concept→relation→concept triples structure into source-concept relation→target-concept binaries thereby making them suitable for FCA. Figure 6 shows the corresponding FCA lattice (i.e. Formal Concept Lattice) that results from this transformation of the corresponding CGs in Fig. 1 from the simple SHU case study. The lattice for the four joined CGs for the expanded example are given by Fig. 7.

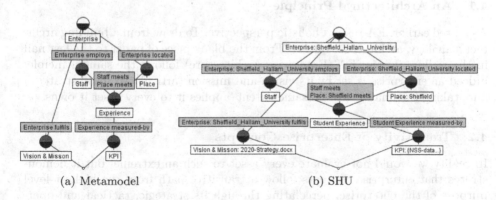

(a) Metamodel   (b) SHU

Fig. 6. Metamodel and SHU, Formal Concept Lattice (FCL) for each

---

[2] https://sourceforge.net/projects/cgfca/.

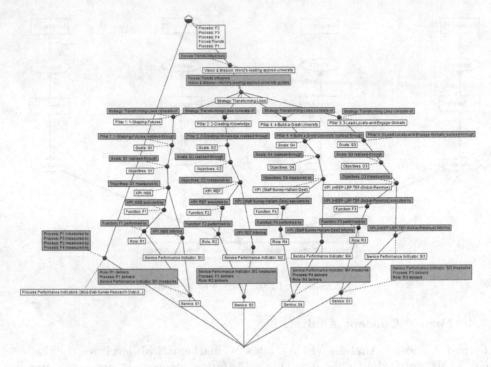

Fig. 7. Modified SHU combined, FCL

4　Iterating Enterprise Concepts from Formal Concepts

We can see that the *infimum* (bottommost) formal concept in Fig. 6 doesn't have its own labels. We will now explore why this is significant.

4.1　An Architectural Principle

As stated earlier, EA takes a holistic perspective. To draw from a building architect's analogy, architecture ranges "From the blank piece of paper to the last nail in the wall." Likewise EA (Enterprise Architecture) follows the same principle; indeed an enterprise is set by its vision and mission (articulated in its strategy) and–taking the analogy to the same extent–applies it to every asset it owns.

4.2　Transitivity of Enterprise Concepts

In reality we would not evaluate every asset to such an extreme, but it demonstrates that enterprise concepts follow a transitive path from the highest level purpose of the enterprise, percolating through its strategic, tactical and operational enterprise concepts as interconnected by their semantic relations to its most specific assets that determine its success. There should be an overall flow from the very top to the very bottom with every concept and relation thus interlinked along the way. In the simple SHU case study, the 'culprit' is the `fulfils` relation in Fig. 1, evident by the upward direction that the arrows point up to `Vision & Mission` from `Enterprise`. All the other arrows point downwards. A formal concept lattice has a *supremum* (topmost) concept and an *infimum* (bottommost) concept. Notably though, the *infimum* has no labels, so what is it's "...to..." enterprise concept? The CGs suggest it's `KPI`, But it's one of the formal concepts above in the lattice. The answer is that the enterprise concepts in Fig. 1 are not all transitive thereby do not concur with the architectural principle. In the expanded SHU case study in Fig. 5 Process delivers a Service, highlighted by the arrow pointing upwards. In fact it should be pointing downwards as `Service` is delivered by a `Process`. It is the `Service` that needs to be changed before the process to ensure that the `Process`' outcome reflects the intended goal [18]. Remember that the metamodel would needed to be validated first, in order to verify any model that is populated from it (as illustrated by the simplified SHU example).

4.3 Correcting the Transitivity

SHU Simple Case Study. Referring to the simple case study (but remembering that SHU is in fact a much more sophisticated enterprise as the more detailed case study identifies), the direction of the arrows around the `fulfils` relation simply need to change direction as stated. This correction is given by Fig. 8, which also shows the `fulfils` relation has become `fulfilled-by`. Although it's a simple renaming in this case, the metamodel (and the SHU model) is fully transitive i.e. architectural. FCA, through *CGFCA* identified this architectural gap. The CGs are conventionally generated by hand, akin to how metamodels and models are developed in many EA software tool environments[3]. As indicated

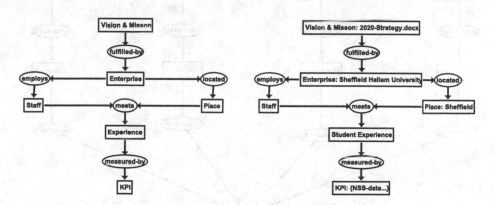

Fig. 8. Corrected metamodel and SHU, in CGs

---

[3] The tools tend to depict the models and metamodels in other notations such as UML (www.uml.org), but this underlying remark still holds true.

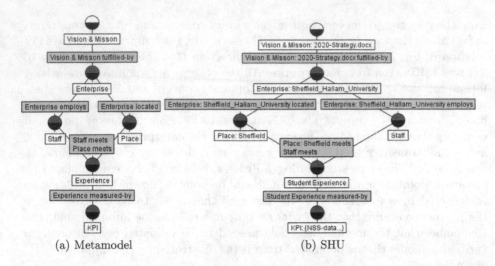

(a) Metamodel   (b) SHU

Fig. 9. Lattices, after correction

earlier, CGs graphical representation serve as a readable, formal design and specification language at a logical level but FCA adds rigour at a mathematical level that allows the formal concepts to be computer generated. The productivity of computers has been applied to the creativity of human thinking—the rationale for conceptual structures (CS).

SHU Extended Case Study. For the expanded SHU example, Fig. 10 shows that two aspects were changed in one of the original four CGs (i.e. part 4) to

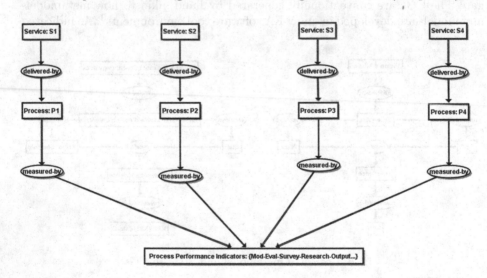

Fig. 10. Corrected SHU 4, in CGs

correct the CG in order to generate a lattice that displays the pathways from the uppermost concept 'Forces and Trends' down to the bottommost concept, PPI.

The first change was the direction of the arrow to represent the correct directional flow between service and process. Then semantic relationship (the second change) is: [Service] → (delivered-by) → [Process] for all the respective referents i.e. S1 to S4, and P1 to P4. This correctly describes the semantic relationship between the concepts and through this the formal concept lattice (Fig. 11) reflects the correct transitivity from *supremum* (topmost) to *infimum* (bottommost) like the simple example, Fig. 9.

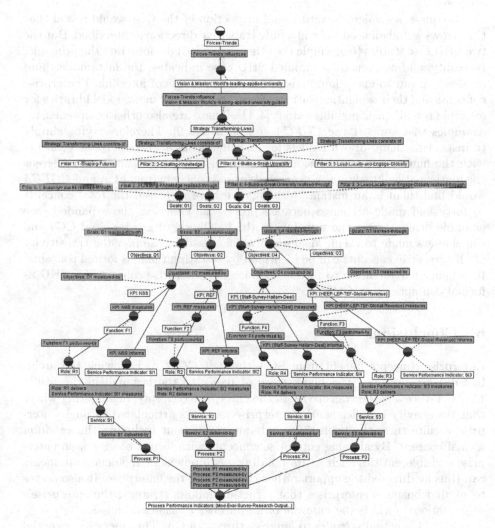

Fig. 11. Corrected SHU combined, FCL

5   Discussion

The significance of the change of arrow direction and relation renaming described above has demonstrated the correct transitivity in relation to the enterprise (SHU) and its strategy. The expanded case example demonstrates a transitive flow that reflects concepts that all connect to the forces and trends which ultimately influence the enterprise strategy. The "...last nail in the wall" *infimum* formal concept PPI brings together all the enterprise concepts to show how they are all (ultimately) measured, hence evaluated and managed for achieving SHU's purpose.

Of course, a straightforward visual inspection of the CGs would reveal that the arrows would all need to be in a fully transitive direction as described. But the two SHU case studies (one simple the other expanded) demonstrate the principle. In reality, and as even the simplified SHU case indicates, the metamodels and models can run to many hundreds and even thousands of interlinked enterprise concepts and their semantic relations. An examination of metamodel libraries for example reveals their possible extent [4, 11]. There are also other comprehensive examples that support the *CGFCA* approach [2, 8, 9]. Therefore trying simply to inspect the hand-drawn models for misalignments in the enterprise concepts with the human eye would become an arduous if not impossible task, whereas the mathematically, computer generated formal concepts from FCA and *CGFCA* would find them in an instant. Although we have increased the total concepts to forty and made a change between process and service, the expanded case example demonstrates the principle in the likelihood of greater-sized CGs and the changes made to them. Meanwhile we can easily sense how the transitivity of all enterprise concepts can be identified by restating them as formal concepts. It is hence our intention to further explore the enterprise concepts for SHU as formal concepts using *CGFCA*.

6   Conclusions

Enterprise concepts benefit from FCA through *CGFCA*. Following the architectural principle of "The blank piece of paper to the last nail in the wall", *CGFCA* discovers the transitivity in the enterprise concepts, highlighting where that transitivity is deficient. For enterprise concepts articulated through enterprise architecture, the transitivity extends throughout including the *infimum* formal concept. By aligning enterprise concepts with formal concepts, an enterprise's visible entities such as its buildings, equipment or financial statements can thus be directed to support rather than hinder the enterprise. It also serves to remind business enterprises that structure follows strategy; the enterprise's organisational form is the *outcome* of its purpose ('vision and mission').

CGFCA is actually *triples* to *binaries* through FCA. This opens its potential to be generalised to other, more widely-used notations that enterprise modellers take advantage of such as UML Class Diagrams that use directed graphs (which are commonly found in EA metamodels). Going even wider, RDFS and OWL from the Semantic Web or any other notation that uses directed triples could benefit too. The experiences from applying *CGFCA* to enterprise metamodels has also raised these additional avenues. Aligning computer productivity with human creativity is a tenet of conceptual structures, and we have shown that FCA in this sense can be brought to bear to make it so.

References

1. Global University Alliance: Industry standards research: the value of applying standards to increase the level of reusability, replication and standardization (2018). http://www.globaluniversityalliance.org/wp-content/uploads/2017/10/Global-University-Alliance-Research-Industry-Standard.pdf
2. Andrews, S., Polovina, S.: A mapping from conceptual graphs to formal concept analysis. In: Andrews, S., Polovina, S., Hill, R., Akhgar, B. (eds.) ICCS 2011. LNCS (LNAI), vol. 6828, pp. 63–76. Springer, Heidelberg (2011). https://doi.org/10.1007/978-3-642-22688-5_5
3. Chandler Jr., A.D.: Strategy and Structure: Chapters in the History of the American Industrial Enterprise. MIT Press, Cambridge (1962)
4. The Open Group: 34. Content metamodel (2011). http://pubs.opengroup.org/architecture/togaf9-doc/arch/chap34.html
5. Gruber, T.R.: A translation approach to portable ontology specifications. Knowl. Acquis. **5**, 199–220 (1993)
6. Hitzler, P., Scharfe, H.: Conceptual Structures in Practice. CRC Press, Boca Raton (2009)
7. Polovina, S.: An introduction to conceptual graphs. In: Priss, U., Polovina, S., Hill, R. (eds.) ICCS-ConceptStruct 2007. LNCS (LNAI), vol. 4604, pp. 1–14. Springer, Heidelberg (2007). https://doi.org/10.1007/978-3-540-73681-3_1
8. Polovina, S., Andrews, S.: CGs to FCA including Peirce's Cuts. Int. J. Concept. Struct. Smart Appl. (IJCSSA) **1**(1), 90–103 (2013)
9. Polovina, S., Scheruhn, H.-J., von Rosing, M.: Modularising the complex metamodels in enterprise systems using conceptual structures. In: Sugumaran, V. (ed.) Developments and Trends in Intelligent Technologies and Smart Systems, pp. 261–283. IGI Global, Hershey (2018). ID: 189437
10. Porter, M.E.: How competitive forces shape strategy. Harv. Bus. Rev. **57**(2), 137–145 (1979). Article on the Positioning School of Strategy
11. LEADing Practice: Meta model reference content #LEAD-ES20021ALL (2018). http://www.leadingpractice.com
12. Roger Sessions: A comparison of the top four enterprise-architecture methodologies (2007). http://msdn.microsoft.com/en-us/library/bb466232.aspx
13. Sowa, J.F.: Conceptual graphs. In: van Harmelen, F., Lifschitz, V., Porter, B. (eds.) Handbook of Knowledge Representation. Foundations of Artificial Intelligence, vol. 3, pp. 213–237. Elsevier, Amsterdam (2008)
14. Sowa, J.F., Zachman, J.A.: Extending and formalizing the framework for information systems architecture. IBM Syst. J. **31**(3), 590–616 (1992)
15. Sowa, J.F.: Conceptual Structures - Information Processing in Mind and Machine. The Systems Programming series. Addison-Wesley, Reading (1984)
16. Sheffield Hallam University: Transforming lives (2017). http://www.shu.ac.uk/strategy

17. von Rosing, M., Fullington, N., Walker, J.: Using the business ontology and enterprise standards to transform three leading organizations. Int. J. Concept. Struct. Smart Appl. (IJCSSA) **4**(1), 71–99 (2016). ID: 171392
18. von Rosing, M., Kirchmer, M.: Focusing business processes on superior value creation: value-oriented process modeling. In: von Rosing, M., Scheer, A.-W., von Scheel, H. (eds.) The Complete Business Process Handbook, pp. 479–496. Morgan Kaufmann, Boston (2015)
19. Zachman, J.A.: John Zachman's concise definition of the Zachman framework. https://www.zachman.com/about-the-zachman-framework
20. Zachman, J.A.: A framework for information systems architecture. IBM Syst. J. **26**(3), 276–292 (1987)

Invited Contributions

Visualizing $\mathcal{ALC}$ Using Concept Diagrams

Gem Stapleton[1(✉)], Aidan Delaney[1,2], Michael Compton[3],
and Peter Chapman[4]

[1] Centre for Secure, Intelligent and Usable Systems,
University of Brighton, Brighton, UK
g.e.stapleton@brighton.ac.uk, aidan@ontologyengineering.org
[2] University of the South Pacific, Suva, Fiji
[3] Hobart, Australia
[4] Edinburgh Napier University, Edinburgh, UK
p.chapman@napier.ac.uk

Abstract. This paper addresses the problem of how to visualize axioms from $\mathcal{ALC}$ using concept diagrams. We establish that 66.4% of OWL axioms defined for ontologies in the Manchester corpus are formulated over $\mathcal{ALC}$, demonstrating the significance of considering how to visualize this relatively simple description logic. Our solution to the problem involves providing a general translation from $\mathcal{ALC}$ axioms into concept diagrams, which is sufficient to establish that all of $\mathcal{ALC}$ can be expressed. However, the translation itself is not designed to give optimally readable diagrams, which is particularly challenging to achieve in the general case. As such, we also improve the translations for a selected category of $\mathcal{ALC}$ axioms, to illustrate that more effective diagrams can be produced.

1 Introduction

Ontology engineering requires a significant skill set as it involves domain modelling and defining axioms using a formal notation, alongside refining and debugging ontologies until the model is seen as accurate and fit-for-purpose. This engineering task can involve many stakeholders, including domain experts who need not be fluent in or, even, familiar with formal notations such as DL or OWL which are typically used by ontology engineers. Communication problems arise as a result. Thus, the use of symbolic notations is a particular obstacle, with this mode of communication potentially leading to inaccurate ontologies being developed or increased time and effort. This is a shortfall because accurate communication of knowledge is necessary for the production of ontologies.

Visualization techniques have been recognized as possible approaches to addressing accessibility problems associated with symbolic notations. Of the various ontology visualization techniques, the majority exploit node-link diagrams (graphs), with OWLViz [13], OntoGraf [2] and CMap [12] being notable examples, but often they are not formalized. These graph-based visualizations exploit the same syntactic element (arrows) to represent both class subsumption and property restrictions. Consequently, the saliency of these two different types

M. Croitoru et al. (Eds.): GKR 2017, LNAI 10775, pp. 99–117, 2018.
https://doi.org/10.1007/978-3-319-78102-0_6

of information is significantly reduced. Similarity theory tells us saliency is an important factor and, in particular, that different syntactic devices should represent different types of information [8]. This is because when visually searching for particular types of information, increasing degrees of similarity between the target syntax (which represents the required information) and distracter syntax (which represents other information) leads to a corresponding increase in the time taken to perform tasks. Another visualization technique is an adaptation of existential graphs, which represent individuals, conjunction and negation using line segments, juxtaposition and closed curves respectively [7]. The resulting notation is essentially a stylized form of first-order logic that uses only $\exists$, $\wedge$ and $\neg$ to make statements and we are of the opinion that usability suffers as a result.

Concept diagrams were introduced for ontology engineering [14] and aid information saliency by avoiding the use of identical (or, even, similar) syntactic types for different informational types: concepts (sometimes called classes) are represented by closed curves and roles (sometimes called properties) by arrows. Figure 1 shows a set of DL axioms, all from $\mathcal{ALC}$, visualized using a single concept diagram; these axioms correspond to a fragment of the SNN ontology [6]. The concept subsumption, concept disjointness and AllValuesFrom-style axioms are represented by curve inclusion, curve disjointness and arrows respectively.

$\mathcal{ALC}$ is an important DL: $\mathcal{ALC}$ axioms form 64.4% of the Manchester corpus [1], which contains over 4500 ontologies comprising nearly 3 million OWL axioms. Whilst the example just given shows how to visualize 25 DL axioms using one

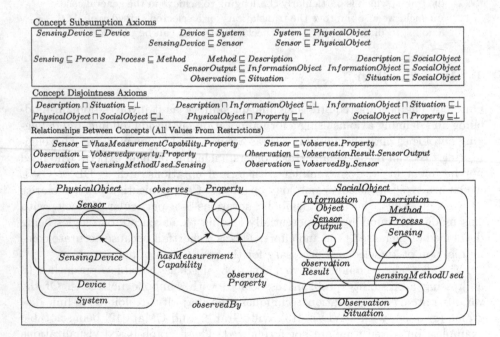

Fig. 1. Description logic axioms converted to a concept diagram.

concept diagram, this paper demonstrates how to translate single DL axioms into diagrams. Our first contribution is to establish concept diagrams equivalent to $\mathcal{ALC}$ concepts. We then go on to establish how to visualize ABox and TBox axioms. Thus, concept diagrams can be used to visualize a significant proportion of the axioms from a large number of ontologies. We also show how to simplify the resulting diagrams into arguably more readable forms.

2   The Description Logic $\mathcal{ALC}$

Readers familiar with the formalization of $\mathcal{ALC}$ may choose to omit this section. In $\mathcal{ALC}$, as with all description logics, axioms are written over a vocabulary comprising a set of individuals, a set of atomic concepts and a set of roles, drawn from the pairwise disjoint sets $\mathcal{O}$, $\mathcal{C}$, and $\mathcal{R}$, respectively. There are two special atomic concepts in $\mathcal{C}$: $\top$ and $\bot$. Individuals, concepts and roles represent elements, sets and binary relations respectively; $\top$ represent Thing (the set containing everything) and $\bot$ represents Nothing (the empty set). The vocabulary is used to form axioms in $\mathcal{ALC}$. Firstly, we define concepts, which are built using atomic concepts and roles along with logical operators and quantifiers.

Definition 1. *The following are **concepts** in $\mathcal{ALC}$:*

1. *Any atomic concept is a concept.*
2. *If C and D are a concepts and R is a role then the following are also (complex) concepts: $(C \sqcap D)$, $(C \sqcup D)$, $\neg C$, $\exists R.C$, and $\forall R.C$.*

In more expressive description logics, other types of concepts can be formed, such as $= nR.C$, which is taken to be the set of things that are related to exactly n things in the 'set' C under the 'relation' R. Moreover, roles can be made more complex, too, such as by forming their composition, $R_1 \circ R_2$, and by taking inverses, R^-. As we are focusing on visualizing axioms drawn from $\mathcal{ALC}$, these more complex constructions are not permitted.

Definition 2. *Given individuals a and b, concepts C and D, and role R the following are **axioms** in $\mathcal{ALC}$: $C(a)$, $R(a,b)$, and $C \sqsubseteq D$. Axioms that involve individuals are **ABox axioms** whereas those which do not are **TBox axioms**.*

We note here that $C \equiv D$ is also sometimes considered an axiom. For the purposes of this paper, we consider $C \equiv D$ to be a pair of subsumption axioms: $C \sqsubseteq D$ and $D \sqsubseteq C$.

Our attention now turns to semantics. Individuals are interpreted as elements, concepts as sets and roles as binary relations.

Definition 3. *An **interpretation** is a pair, $\mathcal{I} = (\Delta^{\mathcal{I}}, \cdot^{\mathcal{I}})$, where*

1. *$\Delta^{\mathcal{I}}$ is a non-empty set, and*
2. *the function $\cdot^{\mathcal{I}}$ maps*
   (a) each individual, a, in $\mathcal{O}$ to an element of $\Delta^{\mathcal{I}}$, that is $a^{\mathcal{I}} \in \Delta^{\mathcal{I}}$,

(b) each concept, C, in $\mathcal{C}$ to a subset of $\triangle^{\mathcal{I}}$, that is $C^{\mathcal{I}} \subseteq \triangle^{\mathcal{I}}$, such that $\top^{\mathcal{I}} = \triangle^{\mathcal{I}}$ and $\bot^{\mathcal{I}} = \emptyset$, and

(c) each role, R, in $\mathcal{R}$ to a binary relation on $\triangle^{\mathcal{I}}$, that is $R^{\mathcal{I}} \subseteq \triangle^{\mathcal{I}} \times \triangle^{\mathcal{I}}$.

The function $\cdot^{\mathcal{I}}$ can then be extended to interpret all concepts as follows:

1. $(C \sqcap D)^{\mathcal{I}} = C^{\mathcal{I}} \cap D^{\mathcal{I}}$,
2. $(C \sqcup D)^{\mathcal{I}} = C^{\mathcal{I}} \cup D^{\mathcal{I}}$,
3. $\neg C^{\mathcal{I}} = \triangle^{\mathcal{I}} \backslash C^{\mathcal{I}}$,
4. $\exists R.C^{\mathcal{I}} = \{x \in \triangle^{\mathcal{I}} : \exists y\, (y \in C^{\mathcal{I}} \wedge (x,y) \in R^{\mathcal{I}})\}$, and
5. $\forall R.C^{\mathcal{I}} = \{x \in \triangle^{\mathcal{I}} : \forall y\, ((y \in \triangle^{\mathcal{I}} \wedge (x,y) \in R^{\mathcal{I}}) \Rightarrow y \in C^{\mathcal{I}})\}$.

Definition 4. *For each axiom, $\mathcal{A}$, an interpretation, $\mathcal{I}$,* **models** $\mathcal{A}$ *under the following conditions:*

1. *If $\mathcal{A} = C(a)$ for some concept C and individual a, $\mathcal{I}$ models $C(a)$ whenever $a^{\mathcal{I}} \in C^{\mathcal{I}}$.*
2. *If $\mathcal{A} = C \sqsubseteq D$ for some concepts C and D then $\mathcal{I}$ models $C \sqsubseteq D$ whenever $C^{\mathcal{I}} \subseteq D^{\mathcal{I}}$.*
3. *If $\mathcal{A} = R(a,b)$ for some role R and individuals a and b then $\mathcal{I}$ models $R(a,b)$ whenever $(a^{\mathcal{I}}, b^{\mathcal{I}}) \in R^{\mathcal{I}}$.*

3   Concept Diagrams

Here we present the formalization of a first-order fragment of the concept diagram logic that is able to express all of $\mathcal{ALC}$. We adapt the formalization given in [23], removing unnecessary second-order, and some first-order, syntax. Firstly, we note that concept diagrams allow the use of inverse roles. So, for every role, R, in $\mathcal{R}$, R^- is a role and we define $R^{-\mathcal{I}} = \{(y,x) : (x,y) \in R^{\mathcal{I}}\}$. Whilst inverse roles are not permitted in $\mathcal{ALC}$, we make use of them in our translation.

An example of a concept diagram is given in Fig. 2. It comprises two *unitary* diagrams, β_1 and β_2; unitary diagrams are extended with additional syntax and are called *class and object property diagrams* in [18]. Each of β_1 and β_2 is enclosed by a *boundary rectangle* which represents the universal set, $\triangle^{\mathcal{I}}$. Each of β_1 and β_2 contain a single *spider*; in β_1 the spider is the graph with two nodes joined by an edge whereas in β_2 the spider comprises just a single node.

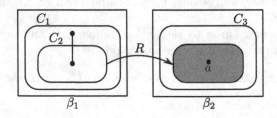

Fig. 2. A concept diagram.

The first spider represents the existence of an anonymous individual whereas the second spider represent the individual a. The labelled (resp. unlabelled) curves represent atomic (resp. anonymous) concepts and shading is used to place upper bounds on set cardinality: in shaded regions, all elements must be represented by spiders. So, in β_2, the only element in the anonymous set is a. Within each unitary diagram, the spatial relationships between the curves and the spiders convey meaning. In β_1, for instance, we can see that $C_2 \sqsubseteq C_1$, through curve inclusion, and the anonymous individual represented by the spider is in C_1. The shading and spider labelled a in β_2 tell us that the only element in the anonymous set is the individual a.

The arrow joining the two unitary diagrams, thus forming a *concept diagram*, asserts that the elements in C_2 (the arrow's source) are, between them, related to *all and only* the elements in the anonymous set represented by the arrow's target which, in turn, is subsumed by C_3. More informally, the arrow tells us that elements in C_2 can only be related to elements in C_3; in $\mathcal{ALC}$, the arrow expresses $C_2 \sqsubseteq \forall R.C_3$. In general, arrows can be sourced and targeted on boundary rectangles, curves and spiders. In addition, arrows can also be dashed to express partial information. In Fig. 2, if the arrow was dashed then the diagram would instead assert that the elements in C_2 are, between them, related to at least all of the elements in the anonymous set represented by the arrow's target.

Our formalization of concept diagrams is at an abstract syntax level. Spiders and closed curves are chosen from countably infinite sets $\mathcal{S}$ and $\mathcal{K}$ respectively; note that these are not closed curves in the mathematical sense. Lastly, arrows represent roles – or, rather, role restriction – are of the form $(s, R, t, \circ)$. Here, s is the arrow's source, R is the arrow's label which is a role or inverse role, t is the target and $\circ$ is either $\rightarrow$ or $--\rightarrow$. As the boundary rectangle in unitary diagrams can be the source or target of an arrow, but is not in $\mathcal{S}$ or $\mathcal{C}$, it will be denoted by $\square$, formally written as $(\square, \beta)$ to identify the diagram, β, in question. Thus, an arrow of the form $((\square, \beta), R, t, \rightarrow)$ indicates that a solid arrow is sourced on the diagram β's boundary rectangle, labelled R with target t.

Definition 5. *A **unitary diagram**, $\beta = (\Sigma, K, \lambda, Z, Z^*, \eta, A)$ has components that are defined as follows.*

1. *$\Sigma = \Sigma(\beta) \subset \mathcal{S}$ is a finite set of spiders.*
2. *$K = K(\beta) \subset \mathcal{K}$ is a finite set of curves.*
3. *$\lambda = \lambda_\beta = \lambda_\Sigma \cup \lambda_K$ is a partial function such that*
   (a) *$\lambda_\Sigma \colon \Sigma \rightarrow \mathcal{O}$ is a partial function that labels spiders with elements $\mathcal{O}$ and*
   (b) *$\lambda_K \colon K \rightarrow \mathcal{C}$ is a partial function that labels curves with elements of $\mathcal{C}$.*
4. *$Z = Z(\beta)$ is a set of **zones** such that $Z \subseteq \{(in, K \backslash in) : in \subseteq K\}$.*
5. *$Z^* = Z^*(\beta) \subseteq Z$ is a set of **shaded zones**.*
6. *$\eta = \eta_\beta \colon \Sigma \rightarrow \mathbb{P}Z \backslash \{\emptyset\}$ is a function that returns the **location** of each spider.*
7. *$A = A(\beta)$ is a finite set of arrows such that for all $(s, R, t, \circ)$ in A, s and t are in $\Sigma \cup K \cup \{(\square, \beta)\}$.*

*A spider or curve that does not map to a label under λ is called **unlabelled**. A set of zones is called a **region**.*

Briefly, β_1 in Fig. 2 has $\Sigma = \{\sigma\}$, $K = \{\kappa_1, \kappa_2\}$, $\lambda(\kappa_1) = C_1$, and $\lambda(\kappa_2) = C_2$. There are three zones (the regions in the plane to which the drawn curves give rise), so $Z = \{(\emptyset, \{\kappa_1, \kappa_2\}), (\{\kappa_1\}, \{\kappa_2\}), (\{\kappa_1, \kappa_2\}, \emptyset)\}$ and none of them are shaded. The function η maps σ to the region $\eta(\sigma) = \{(\{\kappa_1\}, \{\kappa_2\}), (\{\kappa_1, \kappa_2\}, \emptyset)\}$. As β_1 does not contain any arrows (but does contains an arrow source), $A = \emptyset$.

Definition 6. *A **concept diagram** is a tuple, $\mathcal{B} = (\mathcal{D}, A)$, where*

1. *$\mathcal{D}$ is a finite set of unitary diagrams such that for any pair of distinct unitary diagrams, β_1 and β_2, in $\mathcal{D}$ we have $\Sigma(\beta_1) \cap \Sigma(\beta_2) = \emptyset$, and $K(\beta_1) \cap K(\beta_2) = \emptyset$.*
2. *$A = A(\mathcal{B})$ is a finite set of arrows such that for all $(s, R, t, \circ)$ in A, $s, t \in \Sigma(\mathcal{B}) \cup K(\mathcal{B}) \cup (\{\square\} \times \mathcal{D})$ where*

$$\Sigma(\mathcal{B}) = \bigcup_{\beta \in \mathcal{D}} \Sigma(\beta), \quad \text{and} \quad K(\mathcal{B}) = \bigcup_{\beta \in \mathcal{D}} K(\beta)$$

and for all unitary diagrams, β, in $\mathcal{D}$ it is not the case that $s \in \Sigma(\beta) \cup K(\beta) \cup \{(\square, \beta)\}$ and $t \in \Sigma(\beta) \cup K(\beta) \cup \{(\square, \beta)\}$.

The last condition above ensures that arrows in the set $A(\mathcal{B})$ go between different unitary diagrams. This condition can be removed without causing any theoretical problems. It might, however, be counterintuitive if arrows in $A(\mathcal{B})$ simply placed an arrow into one of the unitary parts of the concept diagram. Concept diagrams make use of standard logical connectives to build more complex expressions [23] but these are not needed when focusing on $\mathcal{ALC}$.

Turning our attention to the semantics, the meaning of a unitary diagram is determined by how its individual pieces of syntax are related to each other. We start by translating a unitary diagram into a set of *semantic conditions*. These conditions capture the constraints, provided by the diagram, on the relationships between the represented individuals, concepts, and roles. We start by identifying the elements and sets represented by the labelled spiders and labelled curves. This identification allows us to treat labelled and unlabelled spiders and, respectively, curves, in the same way in our semantic conditions.

Definition 7. *Let β be a unitary diagram and let $\mathcal{I}$ be an interpretation. Let s be a labelled spider and c be a labelled curve in β. We define $s^{\mathcal{I}} = \lambda(s)^{\mathcal{I}}$ and $c^{\mathcal{I}} = \lambda(c)^{\mathcal{I}}$.*

Definition 8. *Let $\mathcal{B} = (\mathcal{D}, A)$ be a concept diagram and let $\mathcal{I} = (\Delta^{\mathcal{I}}, \cdot^{\mathcal{I}})$ be an interpretation, extended so that $(\square, \beta)^{\mathcal{I}} = \Delta^{\mathcal{I}}$, for any β. Then $\mathcal{I}$ is a **model** for $\mathcal{B}$, and $\mathcal{I}$ **satisfies** $\mathcal{B}$, provided there exists an extension of $\mathcal{I}$ to the unlabelled spiders and unlabelled curves in the unitary parts of $\mathcal{B}$, mapping spiders to elements and curves to sets, ensuring the conjunction of the following conditions, called the **semantic conditions**, hold:*

1. For each unitary diagram, β, in $\mathcal{B}$ the following are true.

   (a) **The Curves Condition.** The union of the sets represented by the zones is equal to $\Delta^{\mathcal{I}}$:

   $$\bigcup_{(in,\,out)\in Z(\beta)} (in,\,out)^{\mathcal{I}} = \Delta^{\mathcal{I}}$$

   where

   $$(in,\,out)^{\mathcal{I}} = \bigcap_{\kappa\in in} \kappa^{\mathcal{I}} \cap \bigcap_{\kappa\in out} (\Delta^{\mathcal{I}} \setminus \kappa^{\mathcal{I}}).$$

   (b) **The Shading Condition.** Every shaded zone contains only elements represented by spiders:

   $$\bigwedge_{(in,\,out)\in Z^*(\beta)} (in,\,out)^{\mathcal{I}} \subseteq \{\sigma^{\mathcal{I}} : \sigma \in \Sigma\}.$$

   (c) **The Spiders' Location Condition.** Each spider, σ, represents an element that lies in one of the sets represented by the zones in its location:

   $$\bigwedge_{\sigma\in\Sigma(\beta)} \sigma^{\mathcal{I}} \in \bigcup_{(in,\,out)\in\eta_\beta(\sigma)} (in,\,out)^{\mathcal{I}}.$$

   (d) **The Spiders' Distinctness Condition.** Any two distinct spiders, σ_1 and σ_2, represent distinct elements:

   $$\bigwedge_{\sigma_1,\sigma_2\subseteq\Sigma(\beta)} (\sigma_1 \neq \sigma_2 \Rightarrow \sigma_1^{\mathcal{I}} \neq \sigma_2^{\mathcal{I}}).$$

   (e) **The Arrows Condition.** For each arrow with source s, label R and target t:

   $$\bigwedge_{(s,R,t,\rightarrow)\in A(\beta)} \{y \in \Delta^{\mathcal{I}} : \exists x\,(x \in s^{\mathcal{I}} \wedge (x,y) \in R^{\mathcal{I}})\} = t^{\mathcal{I}} \quad \text{and}$$

   $$\bigwedge_{(s,R,t,\dashrightarrow)\in A(\beta)} \{y \in \Delta^{\mathcal{I}} : \exists x\,(x \in s^{\mathcal{I}} \wedge (x,y) \in R^{\mathcal{I}})\} \supseteq t^{\mathcal{I}}.$$

   where we are treating $s^{\mathcal{I}}$ and $t^{\mathcal{I}}$ as singleton sets, rather than elements, in the cases when s and t, respectively, are spiders.

2. For each arrow, with source s, label R and target t, in $A(B)$, the arrows condition as just given above holds.

An extension of $\mathcal{I}$ that makes the above conditions true is called **appropriate**. Moreover, given a region, r, we define $r^{\mathcal{I}} = \bigcup_{z\in r} z^{\mathcal{I}}$.

4   Building Diagrams for Concepts

Here we provide an inductive construction of concept diagrams for $\mathcal{ALC}$ concepts. The general construction relies on *merging* unitary diagrams. For this operation, as well as other parts of the construction, we rely on diagrams having disjoint curve sets. This reliance is not significant since we can always perform curve substitution, akin to variable substitution in symbolic logics, ensuring that the diagram's components, such as arrow sources and targets and the zones, are updated in the appropriate way; for zones, when substituting κ_1 with κ_2, the zone (in, out) becomes $((in\backslash\{\kappa_1\}) \cup \{\kappa_2\}, out)$ when κ_1 is in in, with the substitution operating similarly when κ_1 is in out. We point out that the construction we give is intended to establish that $\mathcal{ALC}$ axioms *can* all be visualized using concept diagrams. It does not necessarily yield the most effective diagrams, a point to which we return in Sect. 6.

4.1   Merging Diagrams

In order to build diagrams to represent concepts, we need to be able to merge two unitary diagrams that do not contain spiders. An example can be seen in Fig. 3, where β_1 and β_2 are merged into the single diagram $\beta_1 + \beta_2$.

In order to identify the zones of the merged diagram we use the notion of an *expansion* of a region. To illustrate the idea, in Fig. 3, suppose that the curves in β_1 are κ_1 and κ_2 and in β_2 the curves are κ_1' and κ. The region $\{(\{\kappa_1\}, \{\kappa_2\})\}$ in β_1 can be expanded, without changing the set represented, to a four-zone region:

$$\{(\{\kappa_1\}, \{\kappa_2, \kappa_1', \kappa\}), (\{\kappa_1, \kappa_1'\}, \{\kappa_2, \kappa\}), (\{\kappa_1, \kappa\}, \{\kappa_2, \kappa_1'\}), (\{\kappa_1, \kappa_1', \kappa\}, \{\kappa_2\})\}.$$

Definition 9. *Let r be a region and let K be a set of fresh curves (that is no zone in r includes any curve in K). The **expansion** of r given K is the region*

$$\mathcal{EXP}(r, K) = \{(in \cup K', out \cup (K\backslash K')) : (in, out) \in r \wedge K' \subseteq K\}.$$

Lemma 1. *Let r be a region and let K be a set of fresh curves. In any interpretation, $\mathcal{I}$, $r^{\mathcal{I}} = \mathcal{EXP}(r, K)^{\mathcal{I}}$, under any extension of $\mathcal{I}$ mapping curves to sets.*

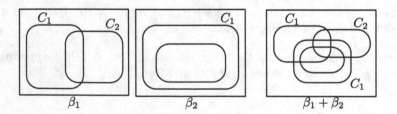

Fig. 3. Merging two diagrams.

When merging two diagrams, we can start the process by expanding their zone sets using the curves in the other diagram. The zones in the merged diagram will be the intersection of these two expansions, thus not including zones that represent empty sets. For instance, considering the four-zone expansion of $(\{\kappa_1\}, \{\kappa_2\})$ given above, the zone $(\{\kappa_1, \kappa\}, \{\kappa_2, \kappa_1'\})$ represents the empty set and is not included in $\beta_1 + \beta_2$. We are now in a position to define how to merge two unitary diagrams that do not contain any spiders.

Definition 10. *Given unitary diagrams* $\beta_1 = (\Sigma_1, K_1, \lambda_1, Z_1, Z_1^*, \eta_1, A_1)$ *and* $\beta_2 = (\Sigma_2, K_2, \lambda_2, Z_2, Z_2^*, \eta_2, A_2)$, *containing no spiders and with disjoint curve sets, their **merger** is a unitary diagram,* $\beta = \beta_1 + \beta_2$, *whose (possibly) non-empty components are:* $K(\beta) = K_1 \cup K_2$, $\lambda_\beta = \lambda_1 \cup \lambda_2$,

$$Z(\beta) = \mathcal{EXP}(Z_1, K_2) \cap \mathcal{EXP}(Z_2, K_1),$$
$$Z^*(\beta) = Z(\beta) \cap (\mathcal{EXP}(Z_1^*, K_2) \cup \mathcal{EXP}(Z_2^*, K_1)),$$

and $A(\beta) = A_1 \cup A_2$.

Lemma 2. *Let* β_1 *and* β_2 *be unitary diagrams with no spiders and disjoint curve sets. Interpretation* $\mathcal{I}$ *models* β_1 *and* β_2 *iff* $\mathcal{I}$ *models* $\beta_1 + \beta_2$.

Proof (Sketch). Follows readily from Lemma 1.

4.2   Translating Concepts into Diagrams

The diagrams we build for concepts express no information, just as the lefthand side and righthand side of an $\mathcal{ALC}$ axiom contain no information when considered in isolation; complex concepts merely describe sets, but do not place any constraints on them (which is done through the use of $\sqsubseteq$ in an axiom, for example). The important feature of diagrams for concepts is that they contain a region that represents the same set as the concept. In what follows, this region is identified diagrammatically by the inclusion of $\times$ as an annotation. The construction is inductive and we begin by defining diagrams for atomic concepts, together with regions that represents the same set as the concept.

Definition 11. *Let* C *be an atomic concept. The **concept diagram for** C, denoted* $\mathcal{DIAG}(C)$, *and the **region for** C, denoted* $\mathcal{REG}(C)$, *are as follows:*

$$\mathcal{DIAG}(C) = \left(\left\{ \boxed{\begin{array}{c} C \\ \bigcirc \times \end{array}} \right\}, \emptyset \right) \quad \text{and} \quad \mathcal{REG}(C) = \{(\{\kappa\}, \emptyset)\}$$

where κ *is the curve labelled* C. *Moreover, the unitary part of* $\mathcal{DIAG}(C)$ *is called the **merging diagram for** C, denoted* $\mathcal{MER}(C)$.

Strictly speaking, the translation of an atomic concept to a diagram returns the *abstract syntax of the concept diagram* but our definition presents a drawing of $\mathcal{DIAG}(C)$ for readability.

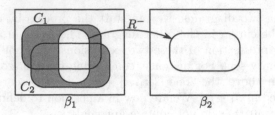

Fig. 4. Translating $\exists R.(C_1 \sqcup C_2)$.

Lemma 3. *Let C be an atomic concept. In any interpretation, $\mathcal{I}$, $C^{\mathcal{I}} = \mathcal{REG}(C)^{\mathcal{I}}$.*

Using this simple base case, we can now build diagrams for complex concepts. In these diagrams, we need to build anonymous concepts using arrows for concepts that involve quantifiers. To facilitate this, we need to add curves inside regions, since arrows cannot be sourced or targeted on the *regions* which represent concepts. To illustrate, Fig. 4 shows a diagram for $\exists R.(C_1 \sqcup C_2)$. Here, the unlabelled curve in β_1 represents the same set as $C_1 \sqcup C_2$. The arrow labelled R^- constructs the set of elements that are related to by some element in $C_1 \sqcup C_2$ and, thus, the unlabelled curve in β_2 represents $\exists R.(C_1 \sqcup C_2)$.

Definition 12. *Let β be a unitary diagram containing no spiders and let r be a region in β. Let κ be a fresh curve. The diagram obtained by **adding** κ inside r, denoted $\beta + (r, \kappa)$ has the same components as β except that the curves are $K(\beta) \cup \{\kappa\}$, the zones are*

$$Z(\beta + (r, \kappa)) = \{(in, out \cup \{\kappa\}) : (in, out) \in Z(\beta) \backslash r\} \cup \mathcal{EXP}(r, \{\kappa\})$$

and the shaded zones are

$$Z^*(\beta + (r, \kappa)) = \{(in, out) \in Z(\beta + (r, \kappa)) : (in \backslash \{\kappa\}, out \backslash \{\kappa\}) \in Z^*(\beta)\} \cup$$
$$\{(in, out \cup \{\kappa\}) : (in, out) \in r\}.$$

Lemma 4. *Let β be a unitary diagram containing no spiders and let r be a region in β. Let κ be a fresh curve. Let $\mathcal{I}$ be an interpretation. Then*

1. *$r^{\mathcal{I}} = \kappa^{\mathcal{I}}$ under any appropriate extension of $\mathcal{I}$ for $\beta + (r, \kappa)$, and*
2. *$\mathcal{I}$ models β iff $\mathcal{I}$ models $\beta + (r, \kappa)$.*

Before we present a definition of the concept diagram for an arbitrary non-atomic concept, we illustrate the key features of the translation by considering $\exists R_1.\neg C_1 \sqcap \forall R_2.(C_2 \sqcup C_3)$. The construction, being inductive, starts by translating the atomic concepts C_1, C_2 and C_3 as in Definition 11. The next stage is to form diagrams for $\neg C_1$ and $C_2 \sqcup C_3$. In fact, the diagram for $\neg C_1$ is the *same* as that for C_1, but the region for $\neg C_1$ differs: it is the complement of the region for C_1. The diagram for $C_2 \sqcup C_3$ is the merger of the diagrams for C_2 and C_3 with,

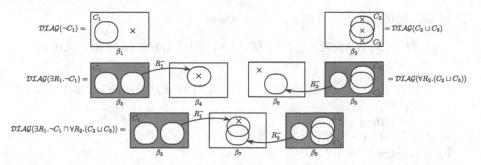

Fig. 5. Translating $\neg C_1$, $C_2 \sqcup C_3$, $\exists R_1.\neg C_1$, $\forall R_2.(C_2 \sqcup C_3)$ and $\exists R_1.\neg C_1 \sqcap \forall R_2.(C_2 \sqcup C_3)$.

roughly speaking, the associated region being the 'union' of the regions for C_2 and C_3. The diagrams for $\neg C_1$ and $C_2 \sqcup C_3$ are β_1 and β_2 respectively, Fig. 5, with their associated regions indicated by ×.

We can now build diagrams for $\exists R_1.\neg C_1$ and $\forall R_2.(C_2 \sqcup C_3)$. Considering $\exists R_1.\neg C_1$, we obtain the concept diagram $(\{\beta_3, \beta_4\}, \{(\kappa_3, R_1^-, \kappa_4, \rightarrow)\})$, where κ_3 and κ_4 are the unlabelled curves in β_3 and β_4 respectively. Here, the diagram which contains the region representing the concept $\exists R_1.\neg C_1$ is β_4; this is the *merging diagram*. For $\forall R_2.(C_2 \sqcup C_3)$, we obtain the concept diagram $(\{\beta_5, \beta_6\}, \{(\kappa_5, R_2^-, \kappa_6, \rightarrow)\})$, where β_6 is the merging diagram. Here, the arrow, together with its source, is used to construct the set of things related to by something not in $C_1 \sqcup C_2$. Thus, the complement of this set – represented by region outside the curve in β_6 – contains exactly the elements that are in $\forall R_2.(C_1 \sqcup C_3)$. The last step is to form a diagram for the entire concept of interest: $\exists R_1.\neg C_1 \sqcap \forall R_2.(C_2 \sqcup C_3)$. We merge β_4 and β_6, leaving β_3 and β_5 unchanged, with the result being the concept diagram $(\{\beta_3, \beta_5, \beta_7\}, \{(\kappa_3, R_1^-, \kappa_4, \rightarrow), (\kappa_5, R_2^-, \kappa_6, \rightarrow)\})$, again with the region representing the entire concept indicated with the inclusion of ×.

Definition 13. *Let C be a non-atomic concept. The **concept diagram for** C, denoted $\mathcal{DIAG}(C)$, the **region for** C, denoted $\mathcal{REG}(C)$, and the **merging diagram for** C, denoted $\mathcal{MER}(C)$, are defined as follows:*

1. *If $C = C_1 \sqcap C_2$ then*
   (a) $\mathcal{MER}(C_1 \sqcap C_2) = \mathcal{MER}(C_1) + \mathcal{MER}(C_2)$,
   (b) $\mathcal{DIAG}(C_1 \sqcap C_2) = (\mathcal{D}, A_1 \cup A_2)$ *where*

$$\mathcal{D} = (\mathcal{D}_1 \setminus \{\mathcal{MER}(C_1)\}) \cup (\mathcal{D}_2 \setminus \{\mathcal{MER}(C_2)\}) \cup \{\mathcal{MER}(C_1 \sqcap C_2)\},$$

   and
   (c) $\mathcal{REG}(C_1 \sqcap C_2) = \mathcal{EXP}(\mathcal{REG}(C_1), K_2) \cap \mathcal{EXP}(\mathcal{REG}(C_2), K_1)$.
2. *If $C = C_1 \sqcup C_2$ then*
   (a) $\mathcal{MER}(C_1 \sqcup C_2) = \mathcal{MER}(C_1) + \mathcal{MER}(C_2)$,

(b) $\mathcal{DIAG}(C_1 \sqcup C_2) = (\mathcal{D}, A_1 \cup A_2)$ where

$$\mathcal{D} = (\mathcal{D}_1 \backslash \{\mathcal{MER}(C_1)\}) \cup (\mathcal{D}_2 \backslash \{\mathcal{MER}(C_2)\}) \cup \{\mathcal{MER}(C_1 \sqcup C_2)\},$$

and

(c) $\mathcal{REG}(C_1 \sqcup C_2) = Z(\mathcal{MER}(C_1 \sqcup C_2)) \cap (\mathcal{EXP}(\mathcal{REG}(C_1), K_2) \cup \mathcal{EXP}(\mathcal{REG}(C_2), K_1)).$

3. If $C = \neg C_1$ then
   (a) $\mathcal{MER}(\neg C_1) = \mathcal{MER}(C_1)$,
   (b) $\mathcal{DIAG}(\neg C_1) = \mathcal{DIAG}(C_1)$ and
   (c) $\mathcal{REG}(\neg C_1) = Z(\mathcal{MER}(C_1)) \backslash \mathcal{REG}(C_1)$.

4. If $C = \exists R.C_1$ then
   (a) $\mathcal{MER}(\exists R.C_1)$ is a unitary diagram containing a fresh curve, κ_t:

   (b) $\mathcal{DIAG}(\exists R.C_1) = (\mathcal{D}, A_1 \cup \{(\kappa_s, R^-, \kappa_t, \rightarrow)\})$ where

   $$\mathcal{D} = (\mathcal{D}_1 \backslash \{\mathcal{MER}(C_1)\}) \cup \{\mathcal{MER}(C_1) + (\mathcal{REG}(C_1), \kappa_s), \mathcal{MER}(\exists R.C_1)\}$$

   and κ_s is a fresh curve, and
   (c) $\mathcal{REG}(\exists R.C_1) = \{(\{\kappa_t\}, \emptyset)\}$.

5. If $C = \forall R.C_1$ then
   (a) $\mathcal{MER}(\forall R.C_1)$ is a unitary diagram containing a fresh curve, κ_t:

   (b) $\mathcal{DIAG}(\forall R.C_1) = (\mathcal{D}, A_1 \cup \{(\kappa_s, R^-, \kappa_t, \rightarrow)\})$ where

   $$\mathcal{D} = (\mathcal{D}_1 \backslash \{\mathcal{MER}(C_1)\}) \cup \\ \{\mathcal{MER}(C_1) + (Z(\mathcal{MER}(C_1)) \backslash \mathcal{REG}(C_1), \kappa_s), \mathcal{MER}(\forall R.C_1)\}$$

   and κ_s is a fresh curve, and
   (c) $\mathcal{REG}(\forall R.C_1) = \{(\emptyset, \{\kappa_t\})\}$.

where $\mathcal{DIAG}(C_1) = (\mathcal{D}_1, A_1)$, $\mathcal{DIAG}(C_2) = (\mathcal{D}_2, A_2)$, and K_1 and K_2 are the sets of curves in $\mathcal{MER}(C_1)$ and $\mathcal{MER}(C_2)$ respectively.

An important property of diagrams for concepts is that they are satisfied in every interpretation. This allows us to readily use them when constructing diagrams for $\mathcal{ALC}$ axioms.

Lemma 5. *Let C be a concept. Then $\mathcal{DIAG}(C)$ is satisfied by all interpretations, that is $\mathcal{DIAG}(C)$ is valid.*

Corollary 1. *Let C be a concept. Then all unitary parts of $\mathcal{DIAG}(C)$ are valid.*

We now establish the crucial result that $\mathcal{REG}(C)$ represents the same set as C.

Theorem 1. *Let C be a concept. For all interpretations, $\mathcal{I}$, $C^{\mathcal{I}} = \mathcal{REG}(C)^{\mathcal{I}}$ under any appropriate extension of $\mathcal{I}$ for $\mathcal{DIAG}(C)$.*

Proof (Sketch). The proof proceeds by induction with the base case provided by Lemma 3. We include the remainder of the proof for the $\exists R.C_1$ and $\forall R.C_1$ cases. In the first of these two cases the curve, κ_t, in $\mathcal{DIAG}(\exists R.C_1)$ that is the target of the arrow represents the image of R^- when its domain is restricted to C_1. Formally, we have

$$\kappa_t^{\mathcal{I}} = \{x \in \Delta^{\mathcal{I}} : \exists y \, (y \in C_1^{\mathcal{I}} \wedge (y,x) \in R^{-\mathcal{I}})\}, \text{by Definition 8}$$

$$= \{x \in \Delta^{\mathcal{I}} : \exists y \, (y \in C_1^{\mathcal{I}} \wedge (x,y) \in R^{\mathcal{I}})\} = \exists R.C_1^{\mathcal{I}}.$$

It is straightforward to verify that $\mathcal{REG}(\exists R.C_1)^{\mathcal{I}} = \{(\{\kappa_t\}, \emptyset)\}^{\mathcal{I}} = \kappa_t^{\mathcal{I}}$ and we are done.

For the $\forall R.C_1$ case, we must show that $\mathcal{REG}(\forall R.C_1) = \{(\emptyset, \{\kappa_t\})\}$ represents the same set as $\forall R.C_1$. Consider $\mathcal{MER}(C_1) + (Z(\mathcal{MER}(C_1), \kappa_s) \backslash \mathcal{REG}(C_1))$. We can show that κ_s, which is the source of the arrow labelled R^-, represents the set $\Delta^{\mathcal{I}} \backslash C_1^{\mathcal{I}}$, using the inductive assumption. Therefore in $\mathcal{MER}(\forall R.C_1)$ the curve, κ_t, which is the target of the arrow labelled R^-, represents the set

$$\kappa_t^{\mathcal{I}} = \{x \in \Delta^{\mathcal{I}} : \exists y \, (y \in \Delta^{\mathcal{I}} \backslash C_1^{\mathcal{I}} \wedge (x,y) \in R^{\mathcal{I}})\} = \exists R.\neg C_1.$$

Thus, $\Delta^{\mathcal{I}} \backslash \kappa_t^{\mathcal{I}}$ contains precisely the elements in $\Delta^{\mathcal{I}}$ that are related only to things in the set $C_1^{\mathcal{I}}$ under $R^{\mathcal{I}}$. More formally,

$$\Delta^{\mathcal{I}} \backslash \kappa_t^{\mathcal{I}} = \left\{ x \in \Delta^{\mathcal{I}} : \forall y \Big((y \in \Delta^{\mathcal{I}} \wedge (x,y) \in R^{\mathcal{I}}) \Rightarrow y \in C_1^{\mathcal{I}} \Big) \right\} = \forall R.C_1^{\mathcal{I}}.$$

Since $\mathcal{REG}(\forall R.C_1) = \{(\emptyset, \{\kappa_t\})\}$, we readily see that $\mathcal{REG}(\forall R.C_1)^{\perp} = \Delta^{\mathcal{I}} \backslash \kappa_t^{\mathcal{I}}$, by definition, and we are done. Hence, in all cases, $C^{\mathcal{I}} = \mathcal{REG}(C)^{\mathcal{I}}$, as required.

5 Visualizing Axioms

In this section we show how to visualize $\mathcal{ALC}$ axioms using concept diagrams.

5.1 ABox Axioms

The Manchester OWL corpus [1] contains over 1.5 million ABox axioms of which 64.3% are in $\mathcal{ALC}$[1]. Using the diagrams constructed for concepts, we are now readily able to establish that all A-box axioms in $\mathcal{ALC}$ can be visualized using concept diagrams. The basic principle for ABox axioms of the form $C(a)$ is to place a spider labelled a in $\mathcal{REG}(C)$. To illustrate, the ABox axiom for $(\exists R_1.\neg C_1 \sqcap \forall R_2.(C_2 \sqcup C_3))(a)$ is visualized in Fig. 6.

---

[1] To count axioms, we used OWL API's DL expressiveness checker. Each axiom is extracted and provided to the OWL API which determines whether the axiom is syntactically in $\mathcal{ALC}$. This approach is somewhat crude, in that some OWL non-$\mathcal{ALC}$ axioms can be reduced to a set of axioms including some in $\mathcal{ALC}$; we count such OWL axioms as not being in $\mathcal{ALC}$. Of the ontologies in the corpus, we could parse 4019. Our counting software is an extension of an existing ontology statistics processing package [11] and can be found at https://github.com/hammar/OntoStats.

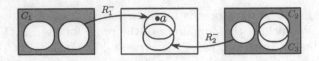

Fig. 6. $(\exists R_1.\neg C_1 \sqcap \forall R_2.(C_2 \sqcup C_3))(a)$.

Definition 14. *The **ABox diagram** for $\mathcal{ALC}$ axiom $C(a)$, denoted $\mathcal{DIAG}$ $(C(a))$, is obtained from $\mathcal{DIAG}(C)$ by adding a spider labelled a to $\mathcal{REG}(C)$ in $\mathcal{MER}(C)$.*

Theorem 2. *Let $C(a)$ be an ABox axiom in $\mathcal{ALC}$ and let $\mathcal{I}$ be an interpretation. $\mathcal{I}$ satisfies $C(a)$ iff $\mathcal{I}$ satisfies $\mathcal{DIAG}(C(a))$.*

Proof. Suppose $\mathcal{I}$ satisfies $C(a)$. Lemma 5 tells us that $\mathcal{I}$ satisfies $\mathcal{DIAG}(C)$. Moreover, Corollary 1 tells us that $\mathcal{I}$ satisfies $\mathcal{MER}(C)$, the unitary part of $\mathcal{DIAG}(C)$ into which the spider, σ say, labelled a has been placed. The only difference between the semantic conditions for $\mathcal{DIAG}(C)$ and $\mathcal{DIAG}(C(a))$ arise from the inclusion of this spider, whereby $\mathcal{DIAG}(C(a))$ asserts:

$$\sigma^{\mathcal{I}} \in \bigcup_{(in,out)\in\mathcal{REG}(C)} (in, out)^{\mathcal{I}} = \mathcal{REG}(C)^{\mathcal{I}} \quad (*).$$

By Theorem 1, $\mathcal{REG}(C)^{\mathcal{I}} = C^{\mathcal{I}}$. Since $\mathcal{I}$ satisfies C, we know that $a^{\mathcal{I}} \in C^{\mathcal{I}}$. By definition, $\sigma^{\mathcal{I}} = a^{\mathcal{I}}$, so (*) is true and we conclude that $\mathcal{I}$ satisfies $\mathcal{DIAG}(C(a))$. The proof for the converse, if $\mathcal{I}$ satisfies $\mathcal{DIAG}(C(a))$ then $\mathcal{I}$ satisfies $C(a)$, is similar. Hence $\mathcal{I}$ satisfies $C(a)$ if and only if $\mathcal{I}$ satisfies $\mathcal{DIAG}(C(a))$.

The remaining ABox case is for axioms of the form $R(a, b)$. These are trivially expressed using concept diagrams:

$$a \bullet\!\!\!-\!-\!\overset{R}{-}\!-\!\bullet b$$

Hence, concept diagrams can express all of $\mathcal{ALC}$'s ABox axioms.

Theorem 3. *All ABox axioms in $\mathcal{ALC}$ can be visualized by a semantically equivalent concept diagram.*

5.2   TBox Axioms

The Manchester OWL corpus [1] contains over 1.3 million TBox axioms of which 66.3% are in $\mathcal{ALC}$. Using the diagrams constructed for concepts, we can establish that all TBox axioms in $\mathcal{ALC}$ can be visualized, although the process is not as straightforward as for ABox axioms. To illustrate, the TBox axiom for $\exists R_1.\neg C_1 \sqsubseteq \forall R_2.(C_2 \sqcup C_3)$ can be seen in Fig. 7 (the diagrams for $\exists R_1.\neg C_1$ and $\forall R_2.(C_2 \sqcup C_3)$ are in Fig. 5). The first step in the construction process is to merge the two merging diagrams for the two sides of the subsumption relationship. This is followed by shading the appropriate zones in order to obtain the

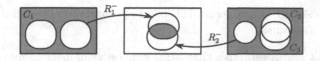

Fig. 7. $\exists R_1.\neg C_1 \sqsubseteq \forall R_2.(C_2 \sqcup C_3)$.

correct subsumption relationship. In this example, there is one zone inside the region for $\exists R_1.\neg C_1$ but not in the region for $\forall R_2.(C_2 \sqcup C_3)$; this zone is shaded to assert that no elements can be in the corresponding set. Formally, the zones which require shading are captured by considering expansions of the regions for $\mathcal{REG}(\exists R_1.\neg C_1)$ and $\mathcal{REG}(\forall R_2.(C_2 \sqcup C_3))$.

Definition 15. *Let $C_1 \sqsubseteq C_2$ be a TBox axiom in $\mathcal{ALC}$ where $\mathcal{DIAG}(C_1) = (\mathcal{D}_1, A_1)$ and $\mathcal{DIAG}(C_2) = (\mathcal{D}_2, A_2)$. The **TBox diagram** for $C_1 \sqsubseteq C_2$, denoted $\mathcal{DIAG}(C_1 \sqsubseteq C_2)$, is obtained from the concept diagram*

$$((\mathcal{D}_1 \backslash \{\mathcal{MER}(C_1)\}) \cup (\mathcal{D}_2 \backslash \{\mathcal{MER}(C_1)\}) \cup \{\mathcal{MER}(C_1) + \mathcal{MER}(C_2)\}, A_1 \cup A_2)$$

by shading the zones in

$$\mathcal{EXP}(\mathcal{REG}(C_1), K(\mathcal{MER}(C_2))) \backslash \mathcal{EXP}(\mathcal{REG}(C_2), K(\mathcal{MER}(C_1)))$$

in $\mathcal{MER}(C_1) + \mathcal{MER}(C_2)$.

Lemma 6. *Let C_1 and C_2 be $\mathcal{ALC}$ concepts. Let $\mathcal{I}$ be an interpretation. The following statements are equivalent.*

(1) $C_1^{\mathcal{I}} \subseteq C_2^{\mathcal{I}}$.
(2) $\mathcal{REG}(C_1)^{\mathcal{I}} \subseteq \mathcal{REG}(C_2)^{\mathcal{I}}$.
(3) $\mathcal{EXP}(\mathcal{REG}(C_1), K(\mathcal{MER}(C_2)))^{\mathcal{I}} \subseteq \mathcal{EXP}(\mathcal{REG}(C_2), K(\mathcal{MER}(C_1)))^{\mathcal{I}}$.

Theorem 4. *Let $C_1 \sqsubseteq C_2$ be a TBox axiom in $\mathcal{ALC}$ and let $\mathcal{I}$ be an interpretation. $\mathcal{I}$ satisfies $C_1 \sqsubseteq C_2$ iff $\mathcal{I}$ satisfies $\mathcal{DIAG}(C_1 \sqsubseteq C_2)$.*

Proof. Suppose that $\mathcal{I}$ satisfies $C_1 \sqsubseteq C_2$. Since $\mathcal{DIAG}(C_1)$ and $\mathcal{DIAG}(C_1)$ are valid, by Lemma 5, we only need to show that $\mathcal{I}$ satisfies the merged unitary diagram $\mathcal{MER}(C_1) + \mathcal{MER}(C_2)$ with the shading added to it as in Definition 15; call this diagram β. First, by Corollary 1, $\mathcal{MER}(C_1)$ and $\mathcal{MER}(C_2)$ are both valid. By Lemma 2, $\mathcal{MER}(C_1) + \mathcal{MER}(C_2)$ is also valid. Therefore, we only need to consider the shading condition for β. This condition reduces to

$$Z^*(\beta)^{\mathcal{I}} = \left(\mathcal{EXP}(\mathcal{REG}(C_1), K(\mathcal{MER}(C_2))) \backslash \mathcal{EXP}(\mathcal{REG}(C_2), K(\mathcal{MER}(C_1))) \right)^{\mathcal{I}} = \emptyset \; (*)$$

since there are no spiders. Now, since $C_1 \sqsubseteq C_2$ is satisfied by $\mathcal{I}$, we know that $C_1^{\mathcal{I}} \subseteq C_2^{\mathcal{I}}$. Lemma 6 tells us, therefore, that

$$\mathcal{EXP}(\mathcal{REG}(C_1), K(\mathcal{MER}(C_2)))^{\mathcal{I}} \subseteq \mathcal{EXP}(\mathcal{REG}(C_2), K(\mathcal{MER}(C_1)))^{\mathcal{I}}$$

from which $(*)$ follows, as required. Thus, $\mathcal{I}$ satisfies $\mathcal{DIAG}(C_1 \sqsubseteq C_2)$. The converse, omitted for space reasons, is similar. Hence $\mathcal{I}$ satisfies $C_1 \sqsubseteq C_2$ iff $\mathcal{I}$ satisfies $\mathcal{DIAG}(C_1 \sqsubseteq C_2)$.

Theorem 5. *All TBox axioms in $\mathcal{ALC}$ can be visualized by a semantically equivalent concept diagram.*

Theorems 3 and 5 establish that $\mathcal{ALC}$ can be visualized using concept diagrams.

6   Improving the General Translations

The translations just defined sometimes return diagrams involving shaded zones. It is possible to simplify these diagrams by removing the shaded zones. An example is given in Fig. 8, where removing shaded zones reduces *clutter*. John et al. [15] defined a clutter score for Euler diagrams (which are concept diagrams that do not include any spiders or arrows): the clutter score for Euler diagram, β, denoted $CS(\beta)$ is

$$CS(\beta) = \sum_{(in,\,out) \in Z(\beta)} |in|.$$

In Fig. 8, the clutter score reduces from 31 to 16 when removing the shaded zones. *All* diagrams arising from TBox axioms involve shading and can be simplified in this way. Moreover, axioms involving quantifiers also give rise to diagrams that include shading.

Lemma 7. *Let A be an $\mathcal{ALC}$ axiom. Removing shaded zones from $\mathcal{DIAG}(A)$ reduces the clutter score.*

It is known that diagrams with a higher clutter score are harder for people to interpret [4] and it has further been shown that Euler diagrams without shading are easier to interpret [5]. Indeed, removing shaded zones makes the resulting diagram exploit spatial relations to assert information, making them well-matched to their semantics [10].

We can also simplify the translation of axioms of the form $C_1 \sqsubseteq \forall R.C_2$, where C_1 and C_2 are arbitrary concepts. For instance, in Fig. 7 the diagram unnaturally uses R_2^- in order to produce a region in β_6 that represents $\forall R_2.(C_2 \sqcup C_3)$. In fact, whilst helpful for a *general translation mechanism*, this construction step can be eliminated, instead making direct use of β_2. An alternative diagram can be seen in Fig. 9. Here, we have added a curve to β_2, Fig. 5, representing a subset of $C_2 \sqcup C_3$. This curve represents the set of all elements that things in $\exists R_1.\neg C_1$ are related to under R_2, though the use of the arrow targeting it. Thus, the diagram expresses $\exists R_1.\neg C_1 \sqsubseteq \forall R_2.(C_2 \sqcup C_3)$. This process readily generalizes

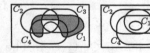

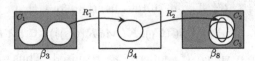

Fig. 8. $C_1 \sqsubseteq C_2 \sqcap C_3 \sqcap C_4$.   **Fig. 9.** $\exists R_1 \neg C_1 \sqsubseteq \forall R_2.(C_2 \sqcup C_3)$.

to axioms of the form $C_1 \sqsubseteq C_2$ where C_2 involves top-level universal quantifiers. We note here that the use of inverse roles for existentially quantified concepts can also be avoided, see [14,21] for examples.

7   Conclusion

This paper shows how to use concept diagrams to visualize $\mathcal{ALC}$ axioms. Our approach was to build diagrams for concepts and then use these diagrams to express ABox and TBox axioms. A substantial proportion of axioms from OWL ontologies are drawn from $\mathcal{ALC}$, establishing that concept diagrams can visualize a significant proportion of ontology axioms that have been developed. We view the contribution in this paper to be an important foundational step towards producing *usable* visualizations of description logic. Whilst our general translation from $\mathcal{ALC}$ may not produce ideal diagrams from a usability perspective, we have demonstrated some improvements can be readily achieved. Further improving the resulting diagrams is a key future ambition. For this, it is likely that extensive empirical studies will be required, to establish how to choose between semantically equivalent, yet syntactically different, concept diagrams. This was started in [3], where diagrams for common styles of axioms where empirically compared to ascertain their relative usability.

There are a number of other exciting avenues for future work. We plan to extend the translations to richer description logics, establishing that most, if not all, ontologies can be visualized using concept diagrams. It will be a particular challenge to produce improved versions of these visualizations, to ensure that the results of translations are most usable. Indeed, we envisage a much more general translation from DL axioms to concept diagrams, which identifies sets of DL axioms that can be translated to single diagrams, as in Fig. 1. We plan to automate the translation process, allowing the results to be readily used in practice. This brings with it substantial diagram drawing and layout problems, building on the body of work on Euler diagram generation [9,16,20,22]. Work towards a theorem prover for concept diagrams has already begun [17], where it has been designed using empirical insights into what constitutes understandable inference rules [19]. Our ultimate vision is to devise a framework that allows concept diagrams to be used for ontology engineering, not merely as a visualization aid, either as a stand-alone notation or fully integrated with existing symbolic approaches.

Acknowledgement. Gem Stapleton was partially funded by a Leverhulme Trust Research Project Grant (RPG-2016-082) for the project entitled Accessible Reasoning with Diagrams.

References

1. Manchester owl corpus. http://owl.cs.manchester.ac.uk/publications/supporting-material/owlcorpus/. Accessed Feb 2014
2. OntoGraf. http://protegewiki.stanford.edu/wiki/OntoGraf. Accessed July 2013
3. Shams, Z., Jamnik, M., Stapleton, G., Sato, Y.: Reasoning with concept diagrams about antipatterns in ontologies. In: Geuvers, H., England, M., Hasan, O., Rabe, F., Teschke, O. (eds.) CICM 2017. LNCS (LNAI), vol. 10383, pp. 255–271. Springer, Cham (2017). https://doi.org/10.1007/978-3-319-62075-6_18
4. Alqadah, M., Stapleton, G., Howse, J., Chapman, P.: Evaluating the impact of clutter in Euler diagrams. In: Dwyer, T., Purchase, H., Delaney, A. (eds.) Diagrams 2014. LNCS (LNAI), vol. 8578, pp. 108–122. Springer, Heidelberg (2014). https://doi.org/10.1007/978-3-662-44043-8_15
5. Chapman, P., Stapleton, G., Rodgers, P., Micallef, L., Blake, A.: Visualizing sets: an empirical comparison of diagram types. In: Dwyer, T., Purchase, H., Delaney, A. (eds.) Diagrams 2014. LNCS (LNAI), vol. 8578, pp. 146–160. Springer, Heidelberg (2014). https://doi.org/10.1007/978-3-662-44043-8_18
6. Compton, M., Barnaghi, P., Bermudez, L., Garcia-Castro, R., Corcho, O., Cox, S., Graybeal, J., Hauswirth, M., Henson, C., Herzog, A., Huang, V., Janowicz, K., Kelsey, W.D., Le Phuoc, D., Lefort, L., Leggieri, M., Neuhaus, H., Nikolov, A., Page, K., Passant, A., Sheth, A., Taylor, K.: The SSN ontology of the W3C semantic sensor network incubator group. Web Semant. Sci. Serv. Agents World Wide Web **17**, 25–32 (2012)
7. Dau, F., Ekland, P.: A diagrammatic reasoning system for the description logic $\mathcal{ALC}$. J. Vis. Lang. Comput. **19**(5), 539–573 (2008)
8. Duncan, J., Humphreys, G.: Visual search and stimulus similarity. Psychol. Rev. **96**, 433–458 (1989)
9. Flower, J., Howse, J.: Generating Euler diagrams. In: Hegarty, M., Meyer, B., Narayanan, N.H. (eds.) Diagrams 2002. LNCS (LNAI), vol. 2317, pp. 61–75. Springer, Heidelberg (2002). https://doi.org/10.1007/3-540-46037-3_6
10. Gurr, C.: Effective diagrammatic communication: syntactic, semantic and pragmatic issues. J. Vis. Lang. Comput. **10**(4), 317–342 (1999)
11. Hammar, K.: Reasoning performance indicators for ontology design patterns. In: 4th Workshop on Ontology and Semantic Web Patterns (2013)
12. Hayes, P., Eskridge, T., Mehrotra, M., Bobrovnikoff, D., Reichherzer, T., Saavedra, R.: COE: tools for collaborative ontology development and reuse. In: Knowledge Capture Conference (2005)
13. Horridge, M.: Owlviz. www.co-ode.org/downloads/owlviz/. Accessed June 2009
14. Howse, J., Stapleton, G., Taylor, K., Chapman, P.: Visualizing ontologies: a case study. In: Aroyo, L., Welty, C., Alani, H., Taylor, J., Bernstein, A., Kagal, L., Noy, N., Blomqvist, E. (eds.) ISWC 2011. LNCS, vol. 7031, pp. 257–272. Springer, Heidelberg (2011). https://doi.org/10.1007/978-3-642-25073-6_17
15. John, C., Fish, A., Howse, J., Taylor, J.: Exploring the notion of 'Clutter' in Euler diagrams. In: Barker-Plummer, D., Cox, R., Swoboda, N. (eds.) Diagrams 2006. LNCS (LNAI), vol. 4045, pp. 267–282. Springer, Heidelberg (2006). https://doi.org/10.1007/11783183_36
16. Riche, N., Dwyer, T.: Untangling Euler diagrams. IEEE Trans. Visual Comput. Graphics **16**(6), 1090–1099 (2010)

17. Shams, Z., Jamnik, M., Stapleton, G., Sato, Y.: Reasoning with concept diagrams about antipatterns. In: 21st International Conference on Logic for Programming, Artificial Intelligence, and Reasoning. pp. 27–42. Kapla Publications in Computing (2017)
18. Shams, Z., Jamnik, M., Stapleton, G., Sato, Y.: Reasoning with concept diagrams about antipatterns in ontologies. In: Geuvers, H., England, M., Hasan, O., Rabe, F., Teschke, O. (eds.) CICM 2017. LNCS (LNAI), vol. 10383, pp. 255–271. Springer, Cham (2017). https://doi.org/10.1007/978-3-319-62075-6_18
19. Shams, Z., Sato, Y., Jamnik, M., Stapleton, G.: Accessible reasoning with diagrams: from cognition to automation. In: 10th International Conference on the Theory and Application of Diagrams. LNCS, vol. 10871. Springer (2018)
20. Simonetto, P., Auber, D., Archambault, D.: Fully automatic visualisation of overlapping sets. Comput. Graphics Forum **28**(3), 967–974 (2009)
21. Stapleton, G., Compton, M., Howse, J.: Visualizing OWL 2 using diagrams. In: IEEE Symposium on Visual Languages and Human-Centric Computing, pp. 245–253. IEEE (2017)
22. Stapleton, G., Flower, J., Rodgers, P., Howse, J.: Automatically drawing Euler diagrams with circles. J. Vis. Lang. Comput. **23**, 163–193 (2012)
23. Stapleton, G., Howse, J., Chapman, P., Delaney, A., Burton, J., Oliver, I.: Formalizing concept diagrams. In: 19th International Conference on Distributed Multimedia Systems, pp. 182–187. KSI (2013)

Graph Theoretical Properties of Logic Based Argumentation Frameworks: Proofs and General Results

Bruno Yun[1], Madalina Croitoru[1(✉)], Srdjan Vesic[2], and Pierre Bisquert[3]

[1] LIRMM, University of Montpellier, Montpellier, France
croitoru@lirmm.fr
[2] CRIL, University of Artois, Arras, France
[3] INRA, Montpellier, France

Abstract. In this paper we extend our first results concerning the characterisation of the graph structure of logic based argumentation graphs with two main classes of findings. First we provide full proofs for the structural results of argumentation graphs built over Datalog± knowledge base composed of facts and negative constraints solely. Second, we also provide some structural properties for the general case of knowledge bases composed of facts, rules and negative constraints.

1 Introduction

We consider existential rules (Calì et al. 2009) as the underlying logical language for the argumentation framework. Starting from an inconsistent existential rule knowledge base (composed of a set of factual knowledge and an ontology stating positive and negative rules about the factual knowledge), using the instantiation of Croitoru and Vesic (2013) we generate the arguments and the attacks corresponding to the knowledge base. The instantiation has been proven to respect rationality desiderata (Amgoud 2014; Caminada and Amgoud 2007) from the argumentation literature and outputs a set of extensions equivalent to the repairs (Lembo et al. 2010; Bienvenu 2012) of the knowledge base (i.e. the maximum w.r.t. inclusion consistent set of factual knowledge).

In Yun et al. (2018) we have given on overview of structural results for simple knowledge bases composed over facts and negative rules only (without any positive rules). In this paper we extend this first study by detailing the results on simple knowledge bases and providing some results that also hold in the general case. More precisely:

– We first consider the case of the knowledge base solely consisting of factual knowledge and negative constraints (expressing fact incompatibility). We fully prove the following structural properties of the argumentation graphs constructed from such knowledge bases: the existence of several duplicates of the same sub-graph, graph-automorphism induced graph symmetries and specific strongly connected component behaviour. We demonstrate how this serves as

© Springer International Publishing AG, part of Springer Nature 2018
M. Croitoru et al. (Eds.): GKR 2017, LNAI 10775, pp. 118–138, 2018.
https://doi.org/10.1007/978-3-319-78102-0_7

a *complete characterisation of argumentation frameworks obtained from such knowledge bases*. We also show that the cf2 semantics (Baroni et al. 2005; Gaggl and Woltran 2013) coincides with the preferred and naive semantics in the case of argumentation frameworks generated from such knowledge bases (without positive rules) that only contain binary negative constraints. Furthermore, we give an example showing that if ternary negative constraints are added, this equivalence no longer holds.

- Second, for the general case of knowledge bases with any number of facts, rules or negative constraints we unveil the following structural properties: the presence of a complete directed sub-graph, the presence of at least one cycle in the graph and, intriguingly, the fact that the cf2 semantics (Baroni et al. 2005; Gaggl and Woltran 2013) is yielding, in this instantiation, potentially inconsistent bases of arguments.

The significance of our results lies in the *graph theoretical structural analysis of a whole family of potentially real world argumentation graphs*. Such results are important to know for software engineers when designing argumentation solvers. For example, when designing SAT inspired solvers (Cerutti et al. 2013), graph symmetries induce as choice the solvers that better perform in the presence of symmetries (Lagniez et al. 2015). Another practical interest lies in benchmark generation (Yun et al. 2017a). Recently argumentation competitions (Thimm et al. 2016) have been held where the benchmarks are generated based on graph theoretical structures known to be difficult for solvers but not confirmed to appear in practice. Revealing real world behaviour could fill this gap and complete the benchmark spectra of instances. Last, it is important to be aware of logic based argumentation graph behaviour in order to keep a realistic expectation of the added value of argumentation in such domains. It is known (Yun et al. 2017b) that even for a modest knowledge base of 7 facts, 2 rules and 1 binary negative constraints, the generated argumentation graph can take gargantuan proportions reaching 383 arguments and 32768 attacks. This paper will help explain, at least partially, these results. For example, the sub-graph duplicate result directly shows the exponential growth of the argumentation graph when facts are added to the knowledge base. Please note that even if the paper of Yun et al. (2017b) deals with benchmark generation of existential rule knowledge bases it is fundamentally different from this work in at least two ways. First, no graph theoretical properties are demonstrated in Yun et al. (2017b). Second, regarding practical added value of Yun et al. (2017b), the authors provide a benchmark generation tool but given the size of the generated graphs it would be difficult to generate the graphs in order to test for their structure. This paper fills this gap and directly provides a variety of graph theoretical properties such graphs enjoy.

In Sect. 2, we recall the basic notions of existential rules and argumentation. In Sect. 3, we show a complete set of structural properties for argumentation frameworks generated from knowledge bases without rules w.r.t. symmetry, strongly connected components and k-copy graphs. Then, the rest of the section deals with structural results for argumentation frameworks generated from general knowledge bases. We show general structural properties such as the absence of self-attacking arguments but also the presence of complete directed sub-graphs. We conclude with a concrete example showing that the cf2 semantics is not suitable for existentially rule instantiated logical argumentation frameworks as it can output sets with inconsistent bases.

2   Background Notions

The existential rules language (Calì et al. 2009) has attracted much interest recently in the Semantic Web and Knowledge Representation community for its suitability for representing knowledge in a distributed context (such as Ontology Based Data Access (OBDA) applications) (Thomazo and Rudolph 2014; Zhang et al. 2016). It is composed of formulae built with the usual quantifiers ($\exists, \forall$) and *only* two connectors: implication ($\rightarrow$) and conjunction ($\wedge$) and is composed of the following elements:

- A *fact* is a ground atom of the form $p(t_1, \ldots, t_k)$ where p is a predicate of arity k and $t_i, i \in [1, \ldots, k]$ constants.
- An existential *rule* is of the form $\forall \vec{X}, \vec{Y} \; H[\vec{X}, \vec{Y}] \rightarrow \exists \vec{Z} C[\vec{Z}, \vec{X}]$ where H (called the hypothesis) and C (called the conclusion) are existentially closed atoms or conjunctions of existentially closed atoms and $\vec{X}, \vec{Y}, \vec{Z}$ their respective vectors of variables. A *rule is applicable* on a set of facts $\mathcal{F}$ iff there exists a homomorphism from the hypothesis of the rule to $\mathcal{F}$. Applying a rule to a set of facts (also called *chase*) consists of adding the set of atoms of the conclusion of the rule to the facts according to the application homomorphism. Different *chase* mechanisms use different simplifications that prevent infinite redundancies (Baget et al. 2011). We use recognisable classes of existential rules where the chase is guaranteed to stop (Baget et al. 2011).
- A *negative constraint* is a particular kind of rule where C is $\bot$ (*absurdum*). Negative constraints can be of any arity (i.e. the number of atoms in C is not bounded). Negative constraints implement weak negation. Please note that negative constraints generalise simple binary conflicts that can easily be translated between the two representations: $\neg p(\boldsymbol{X})$ is transformed into $np(\boldsymbol{X})$ and the negative constraint $p(\boldsymbol{X}) \wedge np(\boldsymbol{X}) \rightarrow \bot$ is added to the rules set.
- A *knowledge base* $\mathcal{K} = (\mathcal{F}, \mathcal{R}, \mathcal{N})$ is composed of a finite set of facts $\mathcal{F}$, a set of rules $\mathcal{R}$ and a set of negative constraints $\mathcal{N}$. We denote by $\mathcal{C}\ell_{\mathcal{R}}^*(\mathcal{F})$ the *closure* of $\mathcal{F}$ by $\mathcal{R}$ (computed by all possible rule $\mathcal{R}$ applications over $\mathcal{F}$ until a fixed point). $\mathcal{C}\ell_{\mathcal{R}}^*(\mathcal{F})$ is said to be $\mathcal{R}$-*consistent* if no negative constraint hypothesis can be deduced. Otherwise, $\mathcal{C}\ell_{\mathcal{R}}^*(\mathcal{F})$ is $\mathcal{R}$-*inconsistent*.
- Given a knowledge base $\mathcal{K} = (\mathcal{F}, \mathcal{R}, \mathcal{N})$, a set of facts $C \subseteq \mathcal{F}$ is called a *minimal conflict* iff C is $\mathcal{R}$-inconsistent and any strict subset $C' \subset C$ of it is $\mathcal{R}$-consistent. The set of all minimal conflict of $\mathcal{K}$ is denoted $conflicts(\mathcal{K})$. If there are no minimal conflicts there are no attacks.

In the OBDA setting rules and constraints act as an ontology used to "access" different data sources. These sources are prone to inconsistencies. As per literature principles, we suppose that the set of rules is compatible with the set of negative constraints, i.e. the union of those two sets is satisfiable (Lembo et al. 2010). This assumption is made because in OBDA we assume that the ontology is believed to be reliable as it is the result of a robust construction by domain experts. However, as data can be large and heterogeneous due to merging and fusion, in the OBDA setting the data is assumed to be the source of inconsistency. This means that by applying the rules on the set of facts, we might violate a constraint. To handle inconsistency, in this paper we use the existential rule instantiation of argumentation frameworks of Croitoru and Vesic (2013):

- An *argument* (Croitoru and Vesic 2013) in $Datalog^{\pm}$ is composed of a minimal (w.r.t. set inclusion) set of facts called *support* and a *conclusion* entailed from the support. The Skolem chase coupled with the use of decidable classes of $Datalog^{\pm}$ ensures the finiteness of the argumentation framework proposed (following from Baget et al. (2011)). Formally, an argument a is a tuple (H, C) with H a non-empty $\mathcal{R}$-consistent subset of $\mathcal{F}$ and C a set of facts:
  - $H \subseteq \mathcal{F}$ and $Cl^*_{\mathcal{R}}(H) \not\models \bot$ *(consistency)*
  - $C \subseteq Cl^*_{\mathcal{R}}(H)$ *(entailment)*
  - $\nexists H' \subset H$ s.t. $C \subseteq Cl^*_{\mathcal{R}}(H')$ *(minimality)*

  The support H of an argument a is denoted by $Supp(a)$ and the conclusion C by $Conc(a)$.
- The *attack* considered is the undermine (Croitoru and Vesic 2013): a attacks b iff the union of the conclusion of a and an element of the support of b are $\mathcal{R}$-inconsistent. Formally, an argument a attacks an argument b denoted by $(a, b) \in \mathcal{C}$ (or $a\mathcal{C}b$) iff $\exists \phi \in Supp(b)$ s.t. $Conc(a) \cup \{\phi\}$ is $\mathcal{R}$-inconsistent. The set of attackers of an argument a is denoted $Att^-(a) = \{a' \mid a'\mathcal{C}a\}$ and the set of arguments attacked by a, $Att^+(a) = \{a' \mid a\mathcal{C}a'\}$.
- An argumentation framework $AS_{\mathcal{K}} = (\mathcal{A}, \mathcal{C})$ is the corresponding AF of $\mathcal{K}$ where $\mathcal{A}$ is the set of arguments and $\mathcal{C}$ is the corresponding attack relation defined above.
- If X is a set of arguments, $Base(X)$ is the union of the supports of the arguments of X: $Base(X) = \bigcup_{x \in X} Supp(x)$.

Example 1. Let us consider the knowledge base $\mathcal{K} = (\mathcal{F}, \mathcal{R}, \mathcal{N})$ with:
$\mathcal{F} = \{a(m), b(m), c(m), d(m)\}$, $\mathcal{R} = \emptyset$ and $\mathcal{N} = \{\forall x(a(x) \wedge b(x) \wedge c(x) \rightarrow \bot)\}$.
The corresponding argumentation framework $AS_{\mathcal{K}}$ is composed of 36 attacks and the following 13 arguments:

- $a0\_0 : (\{a(m)\}, \{a(m)\})$
- $a1\_0 : (\{b(m)\}, \{b(m)\})$
- $a2\_2 : (\{a(m), b(m)\}, \{a(m), b(m)\})$
- $a3\_0 : (\{c(m)\}, \{c(m)\})$
- $a4\_2 : (\{a(m), c(m)\}, \{a(m), c(m)\})$
- $a5\_2 : (\{b(m), c(m)\}, \{b(m), c(m)\})$
- $a6\_0 : (\{d(m)\}, \{d(m)\})$
- $a7\_2 : (\{a(m), d(m)\}, \{a(m), d(m)\})$
- $a8\_2 : (\{b(m), d(m)\}, \{b(m), d(m)\})$
- $a9\_6 : (\{a(m), b(m), d(m)\}, \{a(m), b(m), d(m)\})$

- $a10\_2 : (\{c(m), d(m)\}, \{c(m), d(m)\})$
- $a11\_6 : (\{a(m), c(m), d(m)\}, \{a(m), c(m), d(m)\})$
- $a12\_6 : (\{b(m), c(m), d(m)\}, \{b(m), c(m), d(m)\})$

Please note that the attack is not symmetric, for instance, the argument $a5\_2$ attacks the argument $a0\_0$ but not conversely.

Let us now recall basic argumentation notions (Dung 1995). Let $\mathcal{AS}$ be an argumentation framework, $S \subseteq \mathcal{A}$ and $a \in \mathcal{A}$. We say that:

- S is *conflict-free* iff there exists no arguments $a, b \in S$ such that $(a, b) \in \mathcal{C}$.
- S *defends* a iff for every argument $b \in \mathcal{A}$, if we have $(b, a) \in \mathcal{C}$ then there exists $c \in S$ such that $(c, b) \in \mathcal{C}$.
- S is *admissible* iff it is conflict-free and defends all its arguments.
- S is a *preferred extension* iff it is a maximal (with respect to set inclusion) admissible set.
- S is a *stable extension* iff it is conflict-free and for all $a \in \mathcal{A} \setminus S$, there exists an argument $b \in S$ such that $(b, a) \in \mathcal{C}$.

Example 2 (cont.). There are 3 stable (resp. preferred) extensions in $\mathcal{AS}_\mathcal{K}$:

- $\varepsilon_1 = \{a0\_0, a1\_0, a2\_2, a6\_0, a7\_2, a8\_2, a9\_6\}$
- $\varepsilon_2 = \{a1\_0, a3\_0, a5\_2, a6\_0, a8\_2, a10\_2, a12\_6\}$
- $\varepsilon_3 = \{a0\_0, a3\_0, a4\_2, a6\_0, a7\_2, a10\_2, a11\_6\}$

It was shown in Croitoru and Vesic (2013) that, for existential rules argumentation frameworks, the set of preferred and stable extensions coincide and correspond to the set of maximally consistent sets of facts (repairs).

Example 3 (cont.). The preferred extension ε_1 corresponds to the repair:
$r_1 = \{a(m), b(m), d(m)\}$.
Indeed, we have that $Base(\varepsilon_1) = r_1$.

3   Structural Results

This section is organised as follows. In Sect. 3.1 we first investigate the graph theoretical results of knowledge bases composed solely of facts and negative constraints. Then, in Sect. 3.2 we investigate the general case where rules are also considered in the argumentation framework.

3.1   Results for Simple Knowledge Bases

The graph theoretical results of this subsection are solely looking at the case where the knowledge base is composed of a set of facts and a set of negative constraints defined on these facts. Therefore, at the basis of the results lies the notion of knowledge base minimal conflict. We exhibit three main results:

- The first result deals with conflict induced structural properties. Namely, we characterise *dummy arguments*, arguments that are un-attacked and that do not attack other arguments, and show the repetitious nature of the argumentation graph by introducing the notion of k-copy graph.
- The second result deepens these results and looks into the symmetries of the argumentation graph based on graph auto-morphisms.
- Last, we look into the connectivity of the graph and demonstrate strongly connected components related results.

Please note that these three points will enable us to completely characterise the structural properties of argumentation graphs generated from knowledge bases without positive rules. We begin by introducing the *scope of a negative constraint* which is the set of all sets of facts on which the negative constraint is applicable.

Definition 1. *Let $K = (\mathcal{F}, \mathcal{R}, \mathcal{N})$ be a knowledge base with $\mathcal{R} = \emptyset$ and $N \in \mathcal{N}$ be a negative constraint. We define the scope of the negative constraint N as the set $\mathcal{F}_N = \{X \subseteq \mathcal{F} \mid X$ is minimal with respect to set inclusion such that there is an homomorphism from the body of N to $X\}$.*

Example 4 (cont.). The scope of the negative constraint $N = \forall x(a(x) \wedge b(x) \wedge c(x) \rightarrow \bot$ is $\mathcal{F}_N = \{\{a(m), b(m), c(m)\}\}$.

We show that the number of un-attacked arguments that do not attack other arguments, called *"dummy arguments"*, depends on the number of facts and the scope of all negative constraints.

Proposition 1. *Let $K = (\mathcal{F}, \mathcal{R}, \mathcal{N})$ be an inconsistent knowledge base such that $\mathcal{R} = \emptyset$ and $|\mathcal{F}| = n$. If $AS_K = (\mathcal{A}, \mathcal{C})$ is the corresponding argumentation framework, there are exactly $2^{n-k} - 1$ dummy arguments a in AS_K such $k = |\bigcup_{N \in \mathcal{N}} \bigcup_{X \in \mathcal{F}_N} X|$.*

Proof. Denote $J = \bigcup_{N \in \mathcal{N}} \bigcup_{X \in \mathcal{F}_N} X$.
      Denote $Unn = \{a \in \mathcal{A} \mid Att^-(a) = Att^+(a) = \emptyset\}$.

1. Let us prove that $|Unn| \geq 2^{n-k} - 1$ with $|J| = k$ and $|\mathcal{F}| = n$. The set $J = \bigcup_{N \in \mathcal{N}} \bigcup_{X \in \mathcal{F}_N} X$ corresponds to the set of facts that trigger at least one negative constraint. Thus, every fact that belongs to $E = \mathcal{F} \setminus J$ is not in any conflict. Since $|E| = n - k$ and $\mathcal{R} = \emptyset$, we conclude that there are at least $2^{n-k} - 1$ arguments that have a non empty subset of E as support. These arguments are not attacked and do not attack other arguments as the elements of their supports and conclusions are not in any conflict.
2. Let us prove that $|Unn| \leq 2^{n-k} - 1$ with $|J| = k$ and $|\mathcal{F}| = n$. By means of contradiction, suppose that there is an argument a that do not attack other arguments and that is not attacked but $Supp(a) \nsubseteq E$. It means there exists a negative constraint N such that $(\bigcup_{X \in \mathcal{F}_N} X) \cap Supp(a) \neq \emptyset$ and there is $X \in \mathcal{F}_N$ such that $X \cap Supp(a) \neq \emptyset$. Now, let us consider $Y = X \setminus Supp(a)$. We know that Y is not empty otherwise there is a contradiction with the

consistency of the support of a. Furthermore, Y is $\mathcal{R}$-consistent since $|Y|$ is strictly inferior to the arity of the negative constraint N. Thus, there is an argument $b = (Y, Y)$ such that $(b, a) \in \mathcal{C}$, contradiction.

Example 5 (cont.). We have that $n = 4$ and since we know that $\bigcup_{N \in \mathcal{N}} \bigcup_{X \in \mathcal{F}_N} X = \{a(m), b(m), c(m)\}$, we conclude that there is $2^{4-3} - 1 = 1$ dummy argument. This argument corresponds to $a6\_0 = (\{d(m)\}, \{d(m)\})$.

We now analyse the related behaviour of atoms in the scope of a negative constraint. To do so we introduce the notion of k-copy graph.

A k-copy graph of a graph is another graph that has k times more arguments and each copy a' of a attacks the same arguments as a and possesses the same attackers. Formally:

Definition 2. *Let $\mathcal{AS} = (\mathcal{A}, \mathcal{C})$ be an argumentation framework. We say that the graph of $\mathcal{AS}$ is a k-copy graph of $\mathcal{AS}' = (\mathcal{A}', \mathcal{C}')$ iff:*

- *$|\mathcal{A}| = k * |\mathcal{A}'|$ and there is a surjective function f from $\mathcal{A}$ to $\mathcal{A}'$ such that for every argument $a' \in \mathcal{A}'$, we have $|W_{a'}| = k$, where $W_{a'} = \{a \in \mathcal{A} \mid f(a) = a'\}$.*
- *For all $a, b \in \mathcal{A}$, $(a, b) \in \mathcal{C}$ iff $(f(a), f(b)) \in \mathcal{C}'$.*

Example 6. In Fig. 1, the graph G' (on the right) is a 2-*copy* graph of the graph G (on the left). We have that $W_a = \{a'_1, a'_2\}, W_b = \{b'_1, b'_2\}, W_c = \{c'_1, c'_2\}$.

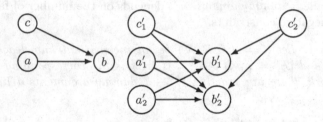

Fig. 1. Representation of a 2-*copy* graph.

If two arguments are the copies of the same argument, then they attack the same arguments and are attacked by the same arguments.

The following proposition shows that if there is a knowledge base $\mathcal{K}$ with no rule and containing k facts that are not in the scope of any negative constraints, then there exists a subgraph of $\mathcal{AS}_\mathcal{K}$ that is a 2^k-*copy* graph of $\mathcal{AS}_{\mathcal{K}'}$ where $\mathcal{K}'$ is the knowledge base with no rules, the same negative constraints as $\mathcal{K}$ and that contains only the facts that are in the scope of at least one negative constraint of $\mathcal{K}$.

This property is important as it shows the behaviour of the instantiation in the case of addition of facts not appearing in any conflict. It shows the structure of the graph and exhibits the exponential growth of the number of arguments w.r.t. these facts.

Proposition 2. *Let* $\mathcal{K} = (\mathcal{F}, \mathcal{R}, \mathcal{N})$ *be a knowledge base with* $\mathcal{R} = \emptyset$.

If $J = \bigcup_{N \in \mathcal{N}} \bigcup_{X \in \mathcal{F}_N} X \neq \emptyset$ *and* $|\mathcal{F} \setminus J| = k$ *then there is a subgraph of* $AS_{\mathcal{K}} = (\mathcal{A}, \mathcal{C})$ *that is a* (2^k)-*copy graph of* $AS_{\mathcal{K}'} = (\mathcal{A}', \mathcal{C}')$ *where* $\mathcal{K}' = (J, \mathcal{R}, \mathcal{N})$ *and* $|\mathcal{A}| = (|\mathcal{A}'| + 1) * 2^k - 1$.

Proof. If $|\mathcal{F} \setminus J| = 0$, then it is obvious that $AS_{\mathcal{K}}$ is a 1-*copy* graph of itself. Suppose now that $|\mathcal{F} \setminus J| > 0$. We denote by $AS_{\mathcal{K}'} = (\mathcal{A}', \mathcal{C}')$ the argumentation framework from the knowledge base $\mathcal{K}' = (J, \mathcal{R}, \mathcal{N})$. Moreover, since $\mathcal{R} = \emptyset$, the arguments can only be of the form (X, X) where X is an $\mathcal{R}$-consistent subset of J. Hence, $|\mathcal{A}'| = |\{X \mid X \text{ is an } \mathcal{R}\text{-consistent subset of } J\}|$.

Now, let us consider $AS_{\mathcal{K}} = (\mathcal{A}, \mathcal{C})$, the argumentation framework corresponding to the knowledge base $\mathcal{K} = (\mathcal{F}, \mathcal{R}, \mathcal{N})$. We show that the subgraph $AS_{\mathcal{K}}'' = (\mathcal{A}'', \mathcal{C}'')$ of $AS_{\mathcal{K}}$ where $\mathcal{A}'' = \{a \in \mathcal{A} \mid Supp(a) \cap J \neq \emptyset\}$ and $\mathcal{C}'' = \mathcal{C}_{|\mathcal{A}''}$ is a $(2^{|\mathcal{F} \setminus J|})$-*copy* graph of $AS_{\mathcal{K}'}$:

- We know that for any set X that is an $\mathcal{R}$-consistent subset of J, $X \cup X'$, where X' is a subset of $\mathcal{F} \setminus J$, is an $\mathcal{R}$-consistent set. Thus $|\mathcal{A}''| = |\{X \cup X' \mid X' \subseteq \mathcal{F} \setminus J \text{ and } X \text{ is an } \mathcal{R}\text{-consistent subset of } J\}|$. Since the number of subsets of $\mathcal{F} \setminus J$ is $2^{|\mathcal{F} \setminus J|}$, then $|\mathcal{A}''| = |\mathcal{A}'| * 2^{|\mathcal{F} \setminus J|}$.
- We denote by f the function from $\mathcal{A}''$ to $\mathcal{A}'$ such that $f(a'') = a'$ iff $Supp(a') = Supp(a'') \cap J$. We now show that this function is surjective. Let a' be an argument of $\mathcal{A}'$ and c an arbitrary element of $\mathcal{F} \setminus J$ (it exists since $|\mathcal{F} \setminus J| > 0$). As mentioned before, we know that $E = Supp(a') \cup \{c\}$ is $\mathcal{R}$-consistent. Therefore $a'' = (E, E)$ is an argument of $\mathcal{A}''$ and $f(a'') = a'$.
- Let $a' \in \mathcal{A}'$ and $W_{a'} = \{a'' \in \mathcal{A}'' \mid f(a'') = a'\}$. For every subset X of $\mathcal{F} \setminus J$, $L = X \cup Supp(a')$, $(L, L) \in W_{a'}$. Since the number of different subsets of $\mathcal{F} \setminus J$ is $2^{|\mathcal{F} \setminus J|}$, we have $|W_{a'}| \geq 2^{|\mathcal{F} \setminus J|}$. Since for every $a'_1, a'_2 \in \mathcal{A}'$, $W_{a'_1} \cap W_{a'_2} = \emptyset$, then for every $a' \in \mathcal{A}', |W_{a'}| = 2^{|\mathcal{F} \setminus J|}$ because $|\mathcal{A}''| = |\mathcal{A}'| * 2^{|\mathcal{F} \setminus J|}$.
- Let $(a''_1, a''_2) \in \mathcal{C}''$, by definition, we have that there exists $\phi \in Supp(a''_2)$ s.t. $Conc(a''_1) \cup \{\phi\}$ is $\mathcal{R}$-inconsistent. Since there are no rules, it is true that $Supp(a''_1) \cup \{\phi\}$ is also $\mathcal{R}$-inconsistent. However, it is clear that this inconsistency cannot come from elements of $\mathcal{F} \setminus J$. Thus, there exists $\phi \in Supp(a''_2) \cap J$ such that $(Supp(a''_1) \cap J) \cup \{\phi\}$ is $\mathcal{R}$-inconsistent. Therefore $(f(a''_1), f(a''_2)) \in \mathcal{C}'$ since $Supp(f(a''_1)) = Supp(a''_1) \cap J$ and $Supp(f(a''_2)) = Supp(a''_2) \cap J$.
- Let $a''_1, a''_2 \in \mathcal{A}''$ such that $(f(a''_1), f(a''_2)) \in \mathcal{C}'$. It means that there exists $\phi \in Supp(f(a''_2))$ s.t. $Conc(f(a''_1)) \cup \{\phi\}$ is $\mathcal{R}$-inconsistent. By definition, we have that $Supp(f(a''_2)) = Supp(a''_2) \cap J$, thus $\phi \in Supp(a''_2)$. Likewise, we have that $Conc(f(a''_1)) = Supp(f(a''_1)) = Supp(a''_1) \cap J = Conc(a''_1) \cap J$. We conclude that $(Conc(a''_1) \cap J) \cup \{\phi\}$ is $\mathcal{R}$-inconsistent. Therefore $Conc(a''_1) \cup \{\phi\}$ is $\mathcal{R}$-inconsistent and $(a''_1, a''_2) \in \mathcal{C}''$.

Finally, we have that $|\mathcal{A}| = |\{X \mid X \text{ is an } \mathcal{R}\text{-consistent subset of } \mathcal{F}\}| = |\{X \mid X \cap J \neq \emptyset \text{ and } X \text{ is an } \mathcal{R}\text{-consistent subset of } \mathcal{F}\} \cup \{X \mid X \subseteq \mathcal{F} \setminus J \text{ and } X \text{ is an } \mathcal{R}\text{-consistent subset of } \mathcal{F}\}| - 1 = |\mathcal{A}'| * 2^{|\mathcal{F} \setminus J|} + 2^{|\mathcal{F} \setminus J|} - 1 = (|\mathcal{A}'| + 1) * 2^{|\mathcal{F} \setminus J|} - 1$. This concludes the proof.

We want to emphasise the result of Proposition 2 as it shows that the addition of "superfluous" facts will increase the size of the argumentation graph by an exponential factor.

Example 7 (cont.). The argumentation framework $\mathcal{AS}_{\mathcal{K}}$ has a subgraph that is a 2-copy graph of $\mathcal{AS}_{\mathcal{K}'}$, where $\mathcal{K}' = (\{a(m),\ b(m),\ c(m)\}, \emptyset, \mathcal{N})$. Indeed, the argumentation framework $\mathcal{AS}_{\mathcal{K}'}$ is composed of the following arguments:

- $a_1 : (\{a(m)\}, \{a(m)\})$
- $a_2 : (\{b(m)\}, \{b(m)\})$
- $a_3 : (\{a(m), b(m)\}, \{a(m), b(m)\})$
- $a_4 : (\{c(m)\}, \{c(m)\})$
- $a_5 : (\{a(m), c(m)\}, \{a(m), c(m)\})$
- $a_6 : (\{b(m), c(m)\}, \{b(m), c(m)\})$

We have that $W_{a_1} = \{a0\_0, a7\_2\}$.

We now focus on detecting symmetries in the graph. Please first note that we have the presence of symmetric arcs in the argumentation framework without rules. It obviously holds that if all negative constraints are binary, then the graph has only symmetric arcs (since the undermining will rely on binary sets). However, if the set of rules is not empty the symmetry no longer holds.

We now explore the link between the instantiation and symmetries in graphs. The next definitions introduce the notions needed to comprehend symmetries, namely, permutations of arguments, orbit of an argument and the cycle notation of a permutation.

Definition 3. *A permutation on a set of elements X is a bijection σ from X to X. Given a permutation σ, the orbit of element $x \in X$ is the set $\mathcal{O}_x = \{x, \sigma(x), \sigma^2(x), \ldots, \sigma^n(x)\}$, with $n \in \{0, 1, \ldots\}$ the minimal integer s.t. $\sigma^{n+1}(x) = x$.*

Definition 4. *Given a permutation σ on X, an orbit $\mathcal{O}$ and an element $x \in \mathcal{O}$, a cycle is a sequence $(x, \sigma(x), \sigma^2(x) \ldots, \sigma^n(x))$, where $n \in \{0, 1, \ldots\}$ is the minimal integer such that $\sigma^{n+1}(x) = x$.*

A permutation can be compactly expressed as a product of cycles corresponding to the orbits of the permutation[1].

Definition 5. *Let $G = (V, E)$ be a graph. A permutation σ on set V is an automorphism of G iff for every two nodes $v_1, v_2 \in V$, we have that $(v_1, v_2) \in E$ iff $(\sigma(v_1), \sigma(v_2)) \in E$.*

The set of automorphisms of a graph, together with the function composition operator, form a group called the automorphism group. The automorphism groups of a graph characterise its symmetries, and are therefore very useful in determining certain of its properties. A subset of a group is called *a generating set of a group* iff every group's element can be expressed as the combination (under group operation) of finitely many elements of the subset and their inverses.

---

[1] In the rest of the paper, and in order to simplify the notation, we omit cycles corresponding to singleton orbits.

Proposition 3. *Let* $\mathcal{AS} = (\mathcal{A}, \mathcal{C})$ *be a k-copy graph of* $\mathcal{AS}' = (\mathcal{A}', \mathcal{C}')$. *For every* $a' \in \mathcal{A}'$, *for every* a_1, a_2 *in* $W_{a'}$, *we have that* (a_1, a_2) *is an automorphism of* $\mathcal{AS}$.

The next proposition shows that if we add nodes (and no arc) to a graph with automorphisms, then the obtained graph also has automorphisms. It is used for showing, in Proposition 5, that a graph constructed on a KB with no rules possesses non trivial automorphisms derived from its subgraph.

Proposition 4. *Let* $G = (V, E)$ *be a graph such that* σ *is an automorphism of* G. *The graph* $G' = (V \cup X, E)$, *where* $X \cap V = \emptyset$, *has the automorphism* σ' *from* $V \cup X$ *to* $V \cup X$:

$$\forall v \in V \cup X, \sigma'(v) = \begin{cases} \sigma(v) & \text{if } v \in V \\ v & \text{if } v \in X \end{cases}$$

Proposition 5. *Let* $\mathcal{K} = (\mathcal{F}, \mathcal{R}, \mathcal{N})$ *with* $\mathcal{R} = \emptyset, J = \bigcup_{N \in \mathcal{N}} \bigcup_{X \in \mathcal{F}_N} X \neq \emptyset, |\mathcal{F} \setminus J| = k, \mathcal{K}' = (J, \mathcal{R}, \mathcal{N})$ *and* $\mathcal{AS}''$ *be a* (2^k)-*copy graph of* $\mathcal{AS}_{\mathcal{K}'} = (\mathcal{A}', \mathcal{C}')$. *If* $\mathcal{AS}''$ *has* k' *automorphisms, then* $\mathcal{AS}_{\mathcal{K}}$ *has at least* k' *automorphisms.*

Proof. From Proposition 2, we know that $\mathcal{AS}_{\mathcal{K}}$ has a subgraph $\mathcal{AS}''_{\mathcal{K}} = (\mathcal{A}'', \mathcal{C}'')$ that is a 2^k-*copy* graph of $\mathcal{AS}_{\mathcal{K}'}$. We first show that every argument a that is in $\mathcal{A} \setminus \mathcal{A}''$ is such that $Att^-(a) = Att^+(a) = \emptyset$. Then we use Proposition 4.

1. We showed in the proof of Proposition 2 that $\mathcal{A}'' = \{a \in \mathcal{A} \mid Supp(a) \cap J \neq \emptyset\}$. Thus, $\mathcal{A} \setminus \mathcal{A}'' = \{a \in \mathcal{A} \mid Supp(a) \subseteq \mathcal{F} \setminus J\}$. Since we have no rules, the arguments in $\mathcal{A} \setminus \mathcal{A}''$ cannot attack other arguments.
2. From Proposition 4, we conclude that there is an automorphism of $\mathcal{AS}_{\mathcal{K}}$ for every automorphism of $\mathcal{AS}''_{\mathcal{K}}$.

Proposition 5 is important as it shows that the graph inherit all of the automorphisms of its subgraph. This will be useful when designing new solvers relying on symmetries.

Example 8 (cont.). Using Propositions 4 and 5, we have that $(a0\_0, a7\_2)$ is an automorphism of $\mathcal{AS}_{\mathcal{K}}$.

We now characterise the connectivity of the graph by showing the structure of the strongly connected components. We first define the impossible set associated to a minimal conflict C as the set containing all the possible subsets of $\mathcal{F}$ that are supersets of at least one subset of C of size $|C - 1|$.

Definition 6. *Let* K *be a knowledge base and* C *a minimal conflict of* $conflicts(\mathcal{K})$. *The impossible set of* C *denoted by* $Imp(C)$ *is* $\{X \subseteq \mathcal{F} \mid X' \subseteq X$ *and* $X' \subseteq C$ *with* $|X'| = |C - 1|\}$.

An argumentation framework is strongly connected if and only if there is a path from any argument a to any argument a'.

Definition 7. *Let $AS = (A, C)$ be an AF. We say that AS is strongly connected iff for every $a, a' \in A'$ such that $a \neq a'$, there is a path from argument a to argument a'.*

Please note that the set of nodes of any arbitrary directed graph can be partitioned such that the subgraphs, induced by each set of nodes, is strongly connected that are called the strongly connected components of this graph. In the rest of this paper, we will denote by $SCC(AS)$, this particular partition of the set of arguments of AS.

In the following proposition, we characterise the structure of the strongly connected components of an argumentation framework obtained from a knowledge base without rules.

Proposition 6. *Let KB be a knowledge base such that $\mathcal{R} = \emptyset$ and $AS_\mathcal{K} = (A, C)$ the corresponding AF. We have that:*

1. *$\{(X_i, X_i)\} \in SCC(AS_\mathcal{K})$ where $X_i \in 2^\mathcal{F} \setminus \bigcup_{C \in conflicts(\mathcal{K})} Imp(C)$*
2. *$(A \setminus \bigcup_i s_i) \in SCC(AS_\mathcal{K})$*

Proof. We split the proof in two parts:

1. Suppose that s_i is not a strongly connected component by itself, it means that there is another argument a such that there is a path from $x_i = \{X_i, X_i\}$ to a and inversely. Let us denote by a_1, the first argument attacked by x_i on a path from x_i to a. By definition, it means that there exists $\phi \in Supp(a_1)$ such that $X_i \cup \{\phi\}$ is $\mathcal{R}$-inconsistent. Since X_i is $\mathcal{R}$-consistent, it means that $X_i \cup \{\phi\}$ is a minimal conflict and that $X_i \in Imp(X_i \cup \{\phi\})$. Then, $Xi \notin 2^\mathcal{F} \setminus \bigcup_{C \in conflicts(\mathcal{K})} Imp(C)$, contradiction.
2. Let a, b be two arguments in $(A \setminus \bigcup_i s_i)$, we show here that there is a path from a to b. From the definitions, we know that a (resp. b) is of the form (X, X) (resp. (X', X')) such that there exists a minimal conflict C (resp. C') and $W \subseteq C$ (resp. $W' \subseteq C'$) with $|W| = |C - 1|$ (resp. $|W'| = |C' - 1|$) and $W \subseteq X$ (resp. $W' \subseteq X'$).
   Let $H = C \setminus X$, $X'' = X' \setminus H$, $W'' \subseteq X''$ with $|W''| = |X'' - 1|$ and $J = H \cup W'' \cup (C' \setminus X')$.
   - If J is $\mathcal{R}$-consistent, we denote by u, the argument (J, J). We have that u belongs to $(A \setminus \bigcup_i s_i)$ because $J = |C' - 1|$ and $J \subseteq C'$. We have that a attacks u and u attacks b.
   - If J is $\mathcal{R}$-inconsistent, it means that there is a minimal conflict $C'' \subseteq J$ such that $C'' \not\subseteq C'$ and $C'' \not\subseteq C$. Let us consider $K, L \subseteq J$ such that $|K| = |L| = |J - 1|$, $H \subseteq K$ and $H \not\subseteq L$. By definition, K and L are $\mathcal{R}$-consistent, thus the arguments $c = (K, K)$ and $d = (L, L)$ exist. We have that a attacks c, c attacks d and d attacks b.

Corollary 1. *Let KB be a knowledge base such that $\mathcal{R} = \emptyset$.*
There are $|2^\mathcal{F} \setminus \bigcup_{C \in conflicts(\mathcal{K})} Imp(C)| + 1$ strongly connected components in $AS_\mathcal{K}$.

Example 9 (cont.). The only minimal conflicts is $C_1 = \{a(m), b(m), c(m)\}$.

We conclude that $2^{\mathcal{F}} \setminus \bigcup_{C \in conflicts(\mathcal{K})} Imp(C) = \{\{a(m)\}, \{b(m)\}, \{c(m)\}, \{d(m)\}, \{a(m), d(m)\}, \{b(m), d(m)\}, \{c(m), d(m)\}\}$ and that there are $7 + 1 = 8$ strongly connected components in $\mathcal{AS}_{\mathcal{K}}$.

We now summarise all the structural properties of the AFs generated from simple knowledge bases using Fig. 2 as an example:

- There is one k-copy graph (encircled in the dashed-line zone).
- The arguments that are not inside the k-copy graph are "dummy arguments" (arguments that are outside the dashed-line zone) and their number can be computed using Proposition 1.
- There is one dense strongly connected component composed of the majority of the arguments (encircled in the grey circle).
- The other strongly connected components are composed of only one argument each (arguments that are outside of the grey circle). The number of strongly connected components can be computed using Corollary 1.

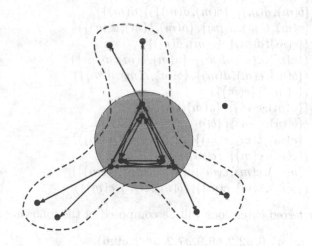

Fig. 2. Structural properties of AFs from simple KBs

Since we deal with strongly connected components, one of the research questions that naturally arise from this is whether or not the cf2 semantics (Baroni et al. 2011; Gaggl and Woltran 2013) is equivalent to the preferred semantics in argumentation graphs generated from knowledge bases without positive rules.

On one hand, it appears that if the set of negative constraints is only composed of binary negative constraints, then the graph only has symmetric arcs. We conclude that since all SCCs are isolated, the cf2 semantics coincides with the naive and preferred semantics.

Proposition 7. *Let KB be a knowledge base such that $\mathcal{R} = \emptyset$, then the cf2 semantics coincides with the preferred and the naive semantics in $AS_{\mathcal{K}}$.*

On the other hand, if we add ternary negative constraints, the cf2 semantics will no longer coincide with the preferred semantics as shown in Example 10.

Example 10. Let $\mathcal{K} = (\mathcal{F}, \mathcal{R}, \mathcal{N})$ be a knowledge base such that $\mathcal{F} = \{a(m), b(m), c(m), d(m), e(m)\}$, $\mathcal{R} = \emptyset$ and $\mathcal{N} = \{\forall x(a(x) \wedge b(x) \wedge c(x) \rightarrow \bot), \forall x(e(x) \wedge d(x) \rightarrow \bot)\}$. The corresponding argumentation framework is composed of 161 attacks and the 20 following arguments:

- $a0\_0 : (\{a(m)\}, \{a(m)\})$
- $a1\_0 : (\{b(m)\}, \{b(m)\})$
- $a2\_2 : (\{a(m), b(m)\}, \{a(m), b(m)\})$
- $a3\_0 : (\{c(m)\}, \{c(m)\})$
- $a4\_2 : (\{a(m), c(m)\}, \{a(m), c(m)\})$
- $a5\_2 : (\{b(m), c(m)\}, \{b(m), c(m)\})$
- $a6\_0 : (\{d(m)\}, \{d(m)\})$
- $a7\_2 : (\{a(m), d(m)\}, \{a(m), d(m)\})$
- $a8\_2 : (\{b(m), d(m)\}, \{b(m), d(m)\})), d(m)\})$
- $a9\_6 : (\{a(m), b(m), d(m)\}, \{a(m), b(m), d(m)\})$
- $a10\_2 : (\{c(m), d(m)\}, \{c(m), d(m)\})$
- $a11\_6 : (\{a(m), c(m), d(m)\}, \{a(m), c(m), d(m)\})$
- $a12\_6 : (\{b(m), c(m), d(m)\}, \{b(m), c(m), d(m)\})$
- $a13\_0 : (\{e(m)\}, \{e(m)\})$
- $a14\_2 : (\{a(m), e(m)\}, \{a(m), e(m)\})$
- $a15\_2 : (\{b(m), e(m)\}, \{b(m), e(m)\})$
- $a16\_6 : (\{a(m), b(m), e(m)\}, \{a(m), b(m), e(m)\})$
- $a17\_2 : (\{c(m), e(m)\}, \{c(m), e(m)\})$
- $a18\_6 : (\{a(m), c(m), e(m)\}, \{a(m), c(m), e(m)\})$
- $a19\_6 : (\{b(m), c(m), e(m)\}, \{b(m), c(m), e(m)\})$

The preferred extensions will be composed of the following sets:

- $\varepsilon_1 = \{a0\_0, a1\_0, a2\_2, a6\_0, a7\_2, a8\_2, a9\_6\}$
- $\varepsilon_2 = \{a0\_0, a3\_0, a4\_2, a6\_0, a7\_2, a10\_2, a11\_6\}$
- $\varepsilon_3 = \{a1\_0, a3\_0, a5\_2, a6\_0, a8\_2, a10\_2, a12\_6\}$
- $\varepsilon_4 = \{a0\_0, a1\_0, a2\_2, a13\_0, a14\_2, a15\_2, a16\_6\}$
- $\varepsilon_5 = \{a0\_0, a3\_0, a4\_2, a13\_0, a14\_2, a17\_2, a18\_6\}$
- $\varepsilon_6 = \{a1\_0, a3\_0, a5\_2, a13\_0, a15\_2, a17\_2, a19\_6\}$

The set of cf2 extensions is the set $\{\varepsilon_1, \varepsilon_2, \varepsilon_3, \varepsilon_4, \varepsilon_5, \varepsilon_6, \varepsilon_7, \varepsilon_8\}$ with:

- $\varepsilon_7 = \{a0\_0, a1\_0, a3\_0, a6\_0, a7\_2, a8\_2, a10\_2\}$
- $\varepsilon_8 = \{a0\_0, a1\_0, a3\_0, a13\_0, a14\_2, a15\_2, a17\_2\}$

3.2   Results for General Knowledge Bases

In this subsection we consider the general case of knowledge bases composed of a set of facts, a set of rules and a set of negative constraints. Unfortunately, as the set of rules can completely change the argumentation framework, general results are much harder to obtain. It is easy to show that the link between the conflict graph (the hyper-graph generated by the negative constraints on the facts, potentially enriched with rules) bares no obvious link to the argumentation graph generated by the corresponding knowledge base. In the Appendix we show that there can be several argumentation frameworks associated with the same minimal conflict graph.

Despite the generality of the problem, we however present three graph theoretical structural results of argumentation graphs:

- First we show general structural properties of the graph: no self-attacking arguments, every argument is defended, having at least one cycle, etc.
- Second we demonstrate the presence of a complete directed sub-graph.
- Third, we show that preferred extensions are included into cf2 extensions but not the other way. Contrary to expectations, we show that the cf2 semantics (originally designed to better handle cycles in graphs) is producing a set of arguments with an inconsistent base.

Let us start by making a few observations on the structure of the argumentation graph. Indeed, it is clear that not any graph can be obtained when constructing arguments from an existential rule knowledge base. First, we remark that an AF generated from a knowledge base $\mathcal{K}$ is always finite. Second, given the definition of an argument, we can also note that there are no self-attacking arguments in our framework:

Proposition 8. *Let $\mathcal{AS} = (\mathcal{A}, \mathcal{C})$ be an argumentation framework s.t. there is an argument $a \in \mathcal{A}$ with $(a, a) \in \mathcal{C}$. There is no Datalog$^{\pm}$ knowledge base $\mathcal{K}$ s.t. $\mathcal{AS}_{\mathcal{K}} = \mathcal{AS}$.*

For every argument in the instantiated argumentation framework, there is a stable (resp. preferred and semi-stable) extension that contains it. Please note that the fact that there are no rejected arguments does not mean that the framework is not expressive as ranking-based semantics may be used to attach more fine-graded acceptability degrees to arguments.

Proposition 9. *Let $\mathcal{AS}_{\mathcal{K}} = (\mathcal{A}, \mathcal{C})$ be the corresponding AF of $\mathcal{K}$. Then, for every argument $a \in \mathcal{A}$, there exists a preferred extension $\varepsilon \in Ext_p(\mathcal{AS}_{\mathcal{K}})$ (resp. semi-stable extension $\varepsilon \in Ext_{ss}(\mathcal{AS}_{\mathcal{K}})$ and stable extension $\varepsilon \in Ext_s(\mathcal{AS}_{\mathcal{K}})$) s.t. $a \in \varepsilon$.*

We now focus on basic observations regarding attacks. First, no knowledge base can generate a framework where an argument a is attacked by an unattacked argument b:

Proposition 10. *Let $\mathcal{AS} = (\mathcal{A}, \mathcal{C})$ be an argumentation framework. If there are two arguments a, b s.t. $(b, a) \in \mathcal{C}$ and there does not exist $c \in \mathcal{A}$ s.t. $(c, b) \in \mathcal{C}$ then there is no Datalog$^{\pm}$ knowledge base $\mathcal{K}$ s.t. $\mathcal{AS}_{\mathcal{K}} = \mathcal{AS}$.*

In the next proposition, we prove the existence of particular arguments associated with a minimal conflict.

Proposition 11. *Let $\mathcal{K}$ be a Datalog$^{\pm}$ knowledge base and $\mathcal{AS}_{\mathcal{K}} = (\mathcal{A}, \mathcal{C})$ the corresponding instantiated AF with C a minimal conflict of $\mathcal{K}$ of size at least 2. If $E, E' \subset C$ such that $|E| = |E'| = |C - 1|$ and $E \neq E'$ then the arguments (E, E) and (E', E') are in $\mathcal{A}$.*

Proof. By definition, we have that E and E' are $\mathcal{R}$-consistent. Suppose that the argument $(E, E) \notin \mathcal{A}$, it means that there is $H \subset E$ and $E \subseteq \mathcal{Cl}_{\mathcal{R}}^{*}(H)$ (minimality). It means that $(C \setminus E) \cup H$ is $\mathcal{R}$-inconsistent and $((C \setminus E) \cup H) \subset C$, contradiction.

If there is at least one minimal conflict C of size at least 2, then there is a cycle[2] in the graph of the instantiated AF:

Proposition 12. *If $\mathcal{K}$ is a Datalog$^{\pm}$ knowledge base and $\mathcal{AS}_{\mathcal{K}} = (\mathcal{A}, \mathcal{C})$ the corresponding instantiated AF with C a minimal conflict of $\mathcal{K}$ of size at least 2 then $\mathcal{AS}_{\mathcal{K}}$ has a cycle.*

Proof. Since there is a minimal conflict of at least size 2 then we know from Proposition 11 that there are two arguments (E, E) and (E', E') in $\mathcal{A}$ such that $E, E' \subset C$, $E \neq E'$ and $|E| = |E'| = |C - 1|$. We have that (E, E) attacks (E', E') and conversely.

Minimal conflicts create a particular structure in the graph of the AF. For every minimal conflict of size n, there is a complete directed subgraph on n nodes (i.e. a subgraph containing n arguments where every argument attacks every argument except itself).

Proposition 13. *Let $\mathcal{K}$ be a knowledge base and $\mathcal{AS}_{\mathcal{K}} = (\mathcal{A}, \mathcal{C})$ the corresponding instantiated AF. For every minimal conflict C of $\mathcal{K}$ s.t. $C \subseteq \mathcal{F}$, there exists a complete directed subgraph of $\mathcal{AS}_{\mathcal{K}}$ with $|C|$ arguments.*

Proof. We consider the case where $|C| > 1$, otherwise it is obvious. Let us consider the set of arguments $\mathcal{A}_C = \{a \in \mathcal{A} | a = (S, S), S \subset C, |S| = |C| - 1\}$. We know that $|\mathcal{A}_C| = |C|$ and for all $a, b \in \mathcal{A}_C$ s.t. $a \neq b$, we have that $(a, b) \in \mathcal{C}$.

Let us now investigate the behaviour of the cf2 semantics on general Datalog$^{\pm}$ argumentation graphs. First, we show that the set of preferred extensions is included in the set of cf2 extensions. We know that in the general case, we have that a stable extension is also a cf2 extension (Gaggl and Woltran 2013; Baroni et al. 2005).

---

[2] We say that a tuple of arguments $(a_1, \ldots, a_n)$ is a cycle if and only if $a_1 \mathcal{C} a_2, \ldots,$ $a_{n-1} \mathcal{C} a_n$ and $a_n \mathcal{C} a_1$.

Proposition 14. *Let $\mathcal{AS}$ be a random AF, we have that $Ext_{st}(\mathcal{AS}) \subseteq Ext_{cf2}(\mathcal{AS})$.*

Furthermore, since we are working in the setting of $Datalog^{\pm}$ argumentation frameworks described in Croitoru and Vesic (2013), a basics result is that the set of preferred extension is equal to the set of stable semantics.

Proposition 15. *Let $\mathcal{AS}$ be a $Datalog^{\pm}$ AF, we have that $Ext_{st}(\mathcal{AS}) = Ext_{pr}(\mathcal{AS})$.*

We thus conclude that the set of preferred extensions is included the set of cf2 extensions for the case of $Datalog^{\pm}$ AFs.

Proposition 16. *Let $\mathcal{AS}$ be a $Datalog^{\pm}$ AF, we have that $Ext_{pr}(\mathcal{AS}) \subseteq Ext_{cf2}(\mathcal{AS})$.*

Note that this result is not true in general (for graphs not generated from $Datalog^{\pm}$ KBs). Moreover, we highlight here that $Ext_{cf2}(\mathcal{AS}) \not\subseteq Ext_{pr}(\mathcal{AS})$ in the $Datalog^{\pm}$ setting by providing the following counter-example.

Example 11. Let us consider the knowledge base $\mathcal{K} = (\mathcal{F}, \mathcal{R}, \mathcal{N})$:
$\mathcal{F} = \{b(m), c(m), d(m), e(m)\}$, $\mathcal{R} = \{\forall x(f(x) \rightarrow b(x))\}$ and $\mathcal{N} = \{\forall x(d(x) \wedge b(x) \wedge c(x) \rightarrow \bot), \forall x(e(x), f(x) \rightarrow \bot)\}$.

The argumentation graph corresponding to this knowledge base is $\mathcal{AS}_{\mathcal{K}} = (\mathcal{A}, \mathcal{C})$ such that $\mathcal{A}$ is composed of:

- $a0\_0 : (\{d(m)\}, \{d(m)\})$
- $a1\_0 : (\{b(m)\}, \{b(m)\})$
- $a1\_1 : (\{b(m)\}, \{f(m)\})$
- $a1\_2 : (\{b(m)\}, \{b(m), f(m)\})$
- $a2\_2 : (\{d(m), b(m)\}, \{d(m), b(m)\})$
- $a2\_4 : (\{d(m), b(m)\}, \{d(m), f(m)\})$
- $a2\_6 : (\{d(m), b(m)\}, \{d(m), b(m), f(m)\})$
- $a3\_0 : (\{c(m)\}, \{c(m)\})$
- $a4\_2 : (\{d(m), c(m)\}, \{d(m), c(m)\})$
- $a5\_2 : (\{b(m), c(m)\}, \{b(m), c(m)\})$
- $a5\_5 : (\{b(m), c(m)\}, \{c(m), f(m)\})$
- $a5\_6 : (\{b(m), c(m)\}, \{b(m), c(m), f(m)\})$
- $a6\_0 : (\{e(m)\}, \{e(m)\})$
- $a7\_2 : (\{d(m), e(m)\}, \{d(m), e(m)\})$
- $a8\_2 : (\{c(m), e(m)\}, \{c(m), e(m)\})$
- $a9\_6 : (\{d(m), c(m), e(m)\}, \{d(m), c(m), e(m)\})$

We have 3 preferred extensions $Ext_{pr} = \{\varepsilon_1, \varepsilon_2, \varepsilon_3\}$ and 4 cf2 extensions $Ext_{cf2} = Ext_{pr} \cup \{\varepsilon_4\}$ with:

- $\varepsilon_1 = \{a0\_0, a1\_0, a1\_1, a1\_2, a2\_2, a2\_4, a2\_6\}$
- $\varepsilon_2 = \{a0\_0, a3\_0, a4\_2, a6\_0, a7\_2, a8\_2, a9\_6\}$

- $\varepsilon_3 = \{a1\_0, a1\_1, a1\_2, a3\_0, a5\_2, a5\_5, a5\_6\}$
- $\varepsilon_4 = \{a0\_0, a1\_0, a1\_1, a1\_2, a2\_4, a3\_0, a5\_5\}$

We showed with Example 11 that the set of cf2 extensions are not included in the set of preferred extensions and thus not equal. Furthermore, contrary to expectations, the cf2 semantics (originally designed to better handle cycles in graphs) is producing a set of arguments with an inconsistent base. Indeed, the set ε_4 contains the arguments a_{0_0} and a_{5_5} which together form an inconsistent base.

4   Discussion

In this paper we investigated the formal structural properties of argumentation graphs generated from $Datalog^{\pm}$ knowledge bases.

We showed that for the case of argumentation frameworks originated from knowledge bases without rules, the *dummy arguments* are the result of facts that are not in the scope of any negative constraints and that their numbers are exponential w.r.t. these facts. Then, we proved that these frameworks possess a particular subgraph called k-copy graph which have symmetries in the form of automorphisms. Moreover, these symmetries can be transferred to the full argumentation framework without loss of generality. Next, we characterised the strong connectivity of the argumentation framework by explaining their structure. Lastly, we showed that the cf2 semantics coincides with the preferred and naive semantics in the case of argumentation frameworks generated from knowledge bases without rules and containing only binary negative constraints.

We then dealt with the case of argumentation frameworks generated from general knowledge bases with rules. We first showed general structural properties of the graph such as the absence of self-attacking arguments, the fact that every argument is defended and the presence of at least one cycle. Second we proved the presence of a complete directed sub-graph associated to each minimal conflict of the knowledge base. Third, we showed that preferred extensions are included into cf2 extensions in this particular instantiation. Last, contrary to expectations, we proved by providing a counter-example that the cf2 semantics (originally designed to better handle cycles in graphs) is producing a set of arguments with an inconsistent base.

The significance of our results lies in the fact that this is the first paper highlighting the *graph theoretical structural analysis of real world argumentation graphs*. We believe that our thorough analysis will enable modellers to understand why and how the changes in the knowledge base can impact the structure of the argumentation framework. What's more, we feel that this paper could be useful for designing faster and better suited solvers for realistic argumentation graphs relying on their inherent structure.

Let us also make a note about the logical language used for instantiating the knowledge bases. Existential rules have been recently intensively investigated for their generalisation with respect to Description Logic fragments. Please note that certain structural results have also been shown to hold in the work of

Arioua et al. (2017b). However, their definition of argument is different from the one used in this paper (as our definition prevents unnecessary repeated arguments). We also note that using argumentation over existential rules has been shown to be of practical interest over existing approaches (Hecham et al. 2017a). Argumentation for handling inconsistency tolerant semantics enhance the human interaction (Arioua and Croitoru 2016), can be used for practical applications in food science (Arioua et al. 2016, 2017a) or allow for alternative computation methods (Yun and Croitoru 2016). Such techniques have been shown to have further implications with respect to human reasoning and bias detection (Bisquert et al. 2016). While the OBDA inspired restriction of inconsistency only coming from the facts could be too strong for certain applications, recently, argumentation inspired approaches that consider defeasible reasoning have been proposed (Hecham et al. 2017b).

Future work will investigate the case of symmetries and strongly connectivity for argumentation graphs from general knowledge bases. Our goal is to obtain a complete characterisation of the argumentation graphs generated from a $Datalog^\pm$ knowledge base.

5   Appendices

We first give the definition of a minimal conflict graph and show that there can be several argumentation frameworks associated with the same minimal conflict graph. This observation is highlighted with an example. We then characterise the arguments and attacks shared by every argumentation frameworks associated with the same minimal conflict graph.

Definition 8. *The minimal conflict graph of an inconsistent knowledge* $\mathcal{K} =$ $(\mathcal{F}, \mathcal{R}, \mathcal{N})$ *is a tuple* $(\mathcal{F}, \mathcal{J}')$, *where* $\mathcal{J}' = conflicts(\mathcal{K})$. *It can be represented with by an hypergraph where elements of* $\mathcal{F}$ *and elements of* $\mathcal{J}'$ *represent nodes and hyper-edges respectively.*

Please note that it is possible that two distinct argumentation frameworks have the same minimal conflict graph.

Example 12. Let $\mathcal{K} = (\mathcal{F}, \mathcal{R}, \mathcal{N})$ and $\mathcal{K}' = (\mathcal{F}, \mathcal{R}', \mathcal{N})$ be two knowledge bases with $\mathcal{F} = \{a(m), b(m)\}$, $\mathcal{R} = \emptyset$, $\mathcal{R}' = \{\forall x(b(x) \rightarrow c(x)\}$ and $\mathcal{N} = \{\forall x(a(x) \wedge b(x) \rightarrow \bot)\}$.

The argumentation framework $\mathcal{AS}_\mathcal{K} = (\mathcal{A}, \mathcal{C})$ is composed of two arguments:

- $a_1 : (\{a(m)\}, a(m))$
- $a_2 : (\{b(m)\}, b(m))$

There are two attacks (a_1, a_2) and (a_2, a_1). However, the argumentation framework $\mathcal{AS}_{\mathcal{K}'} = (\mathcal{A} \cup \{a_3, a_4\}, \mathcal{C}')$ is composed of two more arguments:

- $a_3 : (\{b(m)\}, c(m))$
- $a_4 : (\{b(m)\}, c(m) \wedge b(m))$

We have that $C' = C \cup \{(a_1, a_3), (a_1, a_4), (a_4, a_1)\}$. We remind the reader that these two KBs have the same conflict graph.

Since different argumentation frameworks can have the same conflict graph, it gives us the intuition that there are similarities shared by all these argumentation frameworks.

Definition 9. *The set of consistent subsets of a knowledge base $\mathcal{K}$ is defined as* $consistent(\mathcal{K}) = \{X \subseteq \mathcal{F} \mid \nexists E \in conflicts(\mathcal{K}) \text{ and } E \subseteq X\}$.

Proposition 17. *For every $X, X' \in consistent(\mathcal{K})$ such that there exists $C \in conflicts(\mathcal{K})$ with $C \subseteq X \cup X'$, we have that $(a_1, a_2) \in C$ and $(a_2, a_1) \in C$, where:*

- $a_1 : (X, X)$
- $a_2 : (X', X')$

These arguments and attacks are shared by all the argumentation frameworks sharing the same minimal conflict graph.

References

Amgoud, L.: Postulates for logic-based argumentation systems. Int. J. Approx. Reasoning **55**(9), 2028–2048 (2014)

Arioua, A., Croitoru, M.: A dialectical proof theory for universal acceptance in coherent logic-based argumentation frameworks. In: 22nd European Conference on Artificial Intelligence, ECAI 2016, 29 August–2 September 2016, The Hague, The Netherlands - Including Prestigious Applications of Artificial Intelligence, PAIS 2016, pp. 55–63 (2016)

Arioua, A., Croitoru, M., Buche, P.: DALEK: a tool for dialectical explanations in inconsistent knowledge bases. In: Computational Models of Argument - Proceedings of COMMA 2016, Potsdam, Germany, 12–16 September, 2016, pp. 461–462 (2016)

Arioua, A., Buche, P., Croitoru, M.: Explanatory dialogues with argumentative faculties over inconsistent knowledge bases. Expert Syst. Appl. **80**, 244–262 (2017)

Arioua, A., Croitoru, M., Vesic, S.: Logic-based argumentation with existential rules. Int. J. Approx. Reasoning **90**, 76–106 (2017)

Baget, J.-F., Leclère, M., Mugnier, M.-L., Salvat, E.: On rules with existential variables: walking the decidability line. Artif. Intell. **175**(9–10), 1620–1654 (2011)

Baroni, P., Giacomin, M., Guida, G.: SCC-recursiveness: a general schema for argumentation semantics. Artif. Intell. **168**(1–2), 162–210 (2005)

Baroni, P., Caminada, M., Giacomin, M.: An introduction to argumentation semantics. Knowl. Eng. Rev. **26**(4), 365–410 (2011)

Bienvenu, M.: On the complexity of consistent query answering in the presence of simple ontologies. In: Proceedings of the Twenty-Sixth AAAI Conference on Artificial Intelligence, 22–26 July 2012, Toronto, Ontario, Canada (2012)

Bisquert, P., Croitoru, M., de Saint-Cyr, F.D., Hecham, A.: Substantive irrationality in cognitive systems. In: 22nd European Conference on Artificial Intelligence, ECAI 2016, 29 August–2 September 2016, The Hague, The Netherlands - Including Prestigious Applications of Artificial Intelligence (PAIS 2016), pp. 1642–1643 (2016)

Calì, A., Gottlob, G., Lukasiewicz, T.: A general datalog-based framework for tractable query answering over ontologies. In: Proceedings of the Twenty-Eighth ACM SIGMOD-SIGACT-SIGART Symposium on Principles of Database Systems, PODS 2009, 19 June–1 July 2009, Providence, Rhode Island, USA, pp. 77–86 (2009)

Caminada, M., Amgoud, L.: On the evaluation of argumentation formalisms. Artif. Intell. **171**(5–6), 286–310 (2007)

Cerutti, F., Dunne, P.E., Giacomin, M., Vallati, M.: Computing preferred extensions in abstract argumentation: a SAT-Based approach. In: Theory and Applications of Formal Argumentation-- Second International Workshop, TAFA 2013, Beijing, China, 3–5 August 2013, Revised Selected papers, pp. 176–193 (2013)

Croitoru, M., Vesic, S.: What can argumentation do for inconsistent ontology query answering? In: Liu, W., Subrahmanian, V.S., Wijsen, J. (eds.) SUM 2013. LNCS (LNAI), vol. 8078, pp. 15–29. Springer, Heidelberg (2013). https://doi.org/10.1007/978-3-642-40381-1_2

Dung, P.M.: On the acceptability of arguments and its fundamental role in nonmonotonic reasoning, logic programming and n-Person games. Artif. Intell. **77**(2), 321–358 (1995)

Gaggl, S.A., Woltran, S.: The cf2 argumentation semantics revisited. J. Log. Comput. **23**(5), 925–949 (2013)

Hecham, A., Arioua, A., Stapleton, G., Croitoru, M.: An empirical evaluation of argumentation in explaining inconsistency tolerant query answering. In: 30th International Workshop on Description Logics, DL 2017, Montpellier, France (2017)

Hecham, A., Croitoru, M., Bisquert, P.: Argumentation-based defeasible reasoning for existential rules. In: Proceedings of the 16th Conference on Autonomous Agents and MultiAgent Systems, AAMAS 2017, São Paulo, Brazil, 8–12 May 2017, pp. 1568–1569 (2017)

Lagniez, J.-M., Lonca, E., Mailly, J.-G.: CoQuiAAS: a constraint-based quick abstract argumentation solver. In: 2015 IEEE 27th International Conference on Tools with Artificial Intelligence (ICTAI), pp. 928–935. IEEE (2015)

Lembo, D., Lenzerini, M., Rosati, R., Ruzzi, M., Savo, D.F.: Inconsistency-tolerant semantics for description logics. In: Hitzler, P., Lukasiewicz, T. (eds.) RR 2010. LNCS, vol. 6333, pp. 103–117. Springer, Heidelberg (2010). https://doi.org/10.1007/978-3-642-15918-3_9

Thimm, M., Villata, S., Cerutti, F., Oren, N., Strass, H., Vallati, M.: Summary report of the first international competition on computational models of argumentation. AI Mag. **37**(1), 102 (2016)

Thomazo, M., Rudolph, S.: Mixing materialization and query rewriting for existential rules. In: 21st European Conference on Artificial Intelligence, ECAI 2014, 18–22 August 2014, Prague, Czech Republic - Including Prestigious Applications of Intelligent Systems, PAIS 2014, pp. 897–902 (2014)

Yun, B., Croitoru, M.: An argumentation workflow for reasoning in ontology based data access. In: Computational Models of Argument - Proceedings of COMMA 2016, Potsdam, Germany, 12–16 September 2016, pp. 61–68 (2016)

Yun, B., Croitoru, M., Bisquert, P.: Are ranking semantics sensitive to the notion of core? In: Autonomous Agents and Multiagent Systems - Proceedings of AAMAS 2017, Sao Paulo, Bresil, 8–12 May 2017

Yun, B., Vesic, S., Croitoru, M., Bisquert, P., Thomopoulos, R.: A structural benchmark for logical argumentation frameworks. In: Proceedings of the 20th International Symposium on Intelligent Data Analysis (2017)

Yun, B., Croitoru, M., Vesic, S., Bisquert, P.: A structural benchmark for logical argumentation frameworks. In: Proceedings of the 17th Conference on Autonomous Agents and MultiAgent Systems, AAMAS (2018)

Zhang, H., Zhang, Y., You, J.-H.: Expressive completeness of existential rule languages for ontology-based query answering. In: Proceedings of the Twenty-Fifth International Joint Conference on Artificial Intelligence, IJCAI 2016, New York, NY, USA, 9–15 July 2016, pp. 1330–1337 (2016)

Author Index

Printed in the United States
By Bookmasters